Aeons

Aeons

The Search for the Beginning of Time

MARTIN GORST

FOURTH ESTATE • *London*

For my parents, Roger and Diana Gorst

First published in Great Britain in 2001 by
Fourth Estate
A Division of HarperCollins*Publishers*
77–85 Fulham Palace Road
London W6 8JB
www.4thestate.co.uk

10 9 8 7 6 5 4 3 2 1

A catalogue record for this book is available from
the British Library

ISBN 1-84115-117-3

Typeset by Rowland Phototypesetting Ltd,
Bury St Edmunds, Suffolk
Printed in Great Britain by
The Bath Press Ltd, Bath

Contents

1. In the Beginning

Sometimes the past is nearer than you think. The other day, out of curiosity, I opened my grandmother's Bible, and there, printed alongside the opening verse of Genesis, was the date for the beginning of the world – 4004 BC. This came as a surprise. Not because I didn't know about the date – I did. It was devised by an Irish bishop called James Ussher in the early seventeenth century. No, what surprised me was that anyone should still be proclaiming this as fact as recently as the twentieth century. My grandmother's Bible was printed in 1901. Surely everyone knew by then that the world was far older than this?

I first came across Bishop Ussher's date about fifteen years ago, in a couple of science books written in the 1920s but reprinted in the Thirties as those popular paperbacks you can still pick up for under a pound second-hand. When I went back recently and reread them, what struck me was how seriously they treated what I had regarded as a totally improbable date. Nobody in the twentieth century, I assumed, could really have believed that the world was created – as Ussher maintained – at 6 pm on Saturday 22 October 4004 BC. Yet the passion with which the authors of these books attacked the date gave the appearance of a controversy that, although no longer red-hot, was nevertheless still sizzling. H. G. Wells, writing in 1922, referred to the date as 'this fantastically precise misconception', based upon 'rather arbitrary theological assumptions'. The English geneticist J. B. S. Haldane, writing around the same time, was characteristically more strident: 'we need not pay much attention to clergymen who protest their reverence for Scripture, and yet continue to use, or permit their flocks to use, bibles adorned with the conjectures of an Irish divine'. There's no sign here of the mild amusement with which we view Ussher's date today. These words were the still-glowing embers of a once blazing debate – a debate that had been running since Ussher's dates first appeared in the Bible in the late seventeenth century, and that, in another form, still rages today. At its heart is an inquiry of great significance to both religion and science – the search for the beginning of time.

When did the universe begin? It is one of the simplest, yet one of the most fundamental questions humanity can ask. For the quest to measure the span of the world's existence is more than just an academic pursuit for an abstract figure; it is the point where science and belief merge or clash fatally. Once Ussher's date for the age of the world had been fixed in the pages of the Bible, and science had embarked on its relentless journey of inquiry, then, like a supertanker heading for a reef, collision was inevitable.

Like the famous debate over whether or not the Earth revolved around the sun, the search for the age of the universe was a search for truth. But in this case, the truth took longer to grasp. To begin with, the belief in a short-lived world was so deeply rooted that many of the earliest natural philosophers assumed their studies would confirm the biblical account – the truths of Scripture and the truths of nature, they believed, were one and the same. Only gradually did it become apparent that the world was far older than the Bible said it was. But even after they had accepted this, philosophers struggled to overcome their own preconceived ideas. As they pushed back the frontiers of time they became aware, with growing horror, of its awesome scale. The comforting spans of human experience were eclipsed by ages so enormous that the numbers themselves became inconceivable. Who could imagine a million years, let alone a billion? For some who peered into this abyss, the enormity of what they saw was too startling to reveal to the world; for others – those prepared to publish their results – the conclusions were often too astonishing for the world to comprehend.

As the search broadened from the close reality of the rocks beneath our feet to the distant arena of space above our heads, the evidence itself grew more remote, less graspable. A search that had begun with observations anyone could make relied, by the middle of the twentieth century, on observations of phenomena so small, or again so big, that they were invisible to the population as a whole. The closer science approached the 'truth', the more remote and intangible it became.

When Wells and Haldane attacked Ussher's date in the 1920s, they knew that the universe was much older than the Bible said it was, but they were unable to put a figure on its age. In the last few years, however, science has taken a giant step towards solving this long-running mystery. New and more powerful telescopes, combined with the rapid processing power of modern computers, have enabled astronomers to make measurements that not so long ago they could only dream of. In 1998, for the first time, scientists obtained measurements of all the parameters required to find the universe's age. The search, it seems, may have reached its end.

The fact that there was a search at all, though, is due to Christianity. Two thousand years ago, the idea that the world might have a starting point was inconceivable. Almost all ancient civilisations believed that the universe had existed for ever. From ancient Babylon to early India, the prevailing belief was in an eternal world. Nearly always this concept was combined with the idea of recurring cycles. Instead of having a beginning, time was thought to consist of endless eras, repeated over and over again for eternity. The idea probably grew from the repetitive motion of the heavenly bodies – the daily round of the sun, rising and setting; the monthly waxing and waning of the moon; the annual rotation of the constellations – which all recurred in an endless loop. By 1000 BC, this idea of cyclic time had taken root in the Hindu faith, where the smallest cycle, the Maha Yuga, lasted 4,320,000 years. A thousand of these Maha Yugas made one Kalpa, and two Kalpas completed a single day in the life of Brahma, the chief Hindu deity. After this, the people were reincarnated and the cycle began again.

A similar view was held in ancient Greece. 'Time is infinite and the universe eternal,' proclaimed Aristotle in the fourth century BC. Here, the length of cycle was determined by astronomy. Plato suggested that they lasted the period of time it took for all the planets to return to the same relative positions in their orbits that they had occupied at some earlier time. This span, estimated at 36,000 years, became known as the *great year*, or *magnus annus*, and later became popular in Rome. Here, among the many pagan faiths, astrologers promoted the idea of reincarnation, and the belief that the events of each cycle were repeated in the next. Because time in these cultures had no beginning, the idea of trying to find one simply didn't arise.

Throughout the ancient world there was just one civilisation that didn't subscribe to this cyclic vision of eternity. Jewish scripture, with its story of the Creation, stated clearly that the world had a beginning, a first day when God created the heaven and the earth. By the end of the first century AD, Christianity had adopted this Jewish history as its own, and in the following centuries Christian missionaries spread the idea throughout the Roman world.

In the Beginning

But the idea of everlasting cycles didn't lie easily. At this time, the exact nature of the Creation as described in the Bible was unclear. Just because God had created the world, it didn't necessarily mean that this was the first time it had been created. In a culture dominated by a prevailing belief in an eternal universe, the story of the Creation could easily be viewed as just the beginning of a new cycle, another period of conflagration and rebirth. At the end of the fourth century AD, therefore, there was still no reason to think that time had a beginning. The philosophical bridge between an eternal and a finite world had yet to be crossed.

The idea that time itself had a starting point grew out of a crisis for the Catholic Church. At midnight on 24 August 410 AD, unknown hands opened Rome's Salarian gate from the inside, trumpets blasted, and the Goth army, led by their chieftain Alaric, stormed the city. For three days they ransacked the imperial city of its gold, silver and jewellery – then they left.

When they heard the news, the people of the Roman world – which stretched all around the Mediterranean – reacted with shock. Rome the invincible, the capital of the Empire, the city that had not been captured in 800 years, had fallen. What could have caused such a calamity? The blame was laid on Christianity. For over 700 years, under the protection of the pagan gods, Rome had thrived. Now that the people had abandoned their old gods for Christianity, the gods in turn had abandoned the people. At the same time, the Christian god had failed to protect the city; surely, the people reasoned, this was a false god.

Across the Mediterranean in the North African town of Hippo, the accusations reached the ears of the local bishop, Aurelius Augustinus, known today as St Augustine. Probably the greatest and most influential Christian thinker since St Paul, Augustine realised that the Church faced a crisis. The pagan criticisms had hit their mark, and a wave of doubt had spread through the fledgling Christian community. To quell these doubts, he began writing *The City of God*, a book that would become a landmark in Christian thinking.

His first task was to calm the crisis. By pointing out that the Goths had left the Christian churches of Rome untouched – an act of clemency unprecedented in the history of warfare – he restored the credibility of his faith. But having countered the pagan attack, Augustine continued to expand and develop his arguments in support of Christianity. Over the next three years *The City of God* swelled to twenty-two volumes. Among the difficult questions he strove to clarify was whether or not the Creation truly marked the beginning of time.

For many reasons, Augustine could not accept the prevailing notion that time consisted of everlasting cycles. For a start, if life was predestined, comprising the repetition of events of a previous age, it denied the uniqueness of Jesus Christ. Second, if all one's actions could be put down to predestination, there would be no inducement to follow the good life prescribed by Christianity. And third, he found the popular notion that these cycles repeated themselves absurd. Were we expected to believe that:

> just as in this age the philosopher Plato sat in the city of Athens and in the school called Academy teaching his pupils, so also through countless ages of the past at intervals both the same Plato and the same city and the same school and the same pupils have been repeated, as they are destined to be repeated through countless ages of the future. God forbid, I say, that we should swallow such nonsense!

In Christian philosophy, he maintained, the world could not be eternal. Pagan religions that adhered to cosmic cycles were

In the Beginning

'those argumentations whereby the infidel seeks to undermine our simple faith, dragging us from the straight road and compelling us to walk with him on the wheel'. The creation of the world as described in the Bible, Augustine explained, was not the recreation of a previous world, but the absolute beginning of everything. Before the world was created, time did not exist. God crated the *world* and *time* together. In other words, the world was not created at some moment in time, but simultaneously with it. In Christian philosophy, therefore, the question: 'What happened before God created the universe?' had no meaning. The moment of creation *was* the beginning of time. There still remained the question of what God did *before* he made the heaven and earth. Sadly, Augustine did not answer this question, as is commonly reported, by saying that 'He created Hell, for people who ask those kind of questions.' He simply said that God 'did nothing whatsoever'.

With the success of *The City of God*, and the triumph of Christianity, the old notion of an eternal universe faded from European thought. As Christianity spread through Europe, and later out into the wider world, Augustine's great idea travelled with it. Instead of endless aeons stretching back for eternity, the world now had a starting point and, tantalisingly, the Bible even gave clues that its date could be measured. And so began the search for the beginning of time.

2. The Bishop and the Book

As the morning of 30 January 1649 dawned over London it brought with it a nervous buzz of anticipation. The previous day, workmen had erected a scaffold outside the Banqueting Hall in Whitehall and rumours of the imminent execution of the king had spread like wildfire through the city. In Charing Cross a group of servants and gentlemen climbed up to the frost-covered leads of Lady Peterborough's house, from where they had a clear view of the scaffold and the expectant crowd surrounding it. Numbed by the bitter cold, and a dread of what they were about to witness, the men stood transfixed on the roof, waiting to see what would happen next.

Among the guests at Lady Peterborough's that day was an elderly Irish bishop, James Ussher. Tall, with long hair, and a small grey beard neatly trimmed in the fashion of the day like the king's, Ussher brought a gentle dignity to the house. Unlike the other visitors, he was reluctant to climb up to the roof to watch the gruesome spectacle, for he had once been Charles I's chaplain, and had loyally accompanied the king through the tumult of the recent civil war. Now, at the age of sixty-nine, Ussher was virtually penniless, and for support relied largely on Lady Peterborough's benevolence. Thanks to her kindness, he was at last close to completing his life's work, *The Annals of the World*, a comprehensive history of ancient times, starting with the day of Creation and finishing

The Bishop and the Book

JACOBUS USSERIUS, ARCHIEPISCOPUS ARMACHANUS,
TOTIUS HIBERNIÆ PRIMAS.
London Printed for Nath: Ranew and Jonat: Robinson at the Kings Armes, in S.t Pauls Church yard 1677

Archbishop James Ussher.

in AD 70. It was an immense undertaking and an impressive piece of scholarship, drawing a linear thread from thousands of ancient books and manuscripts written in a multitude of different languages. It had taken over twenty years to write, but at last the first volumes were finished and would be published in London the following summer. Although the completed work stretches to two thousand pages of densely printed Latin, Ussher's fame rests on a single paragraph featuring a single date – Saturday 22 October 4004 BC. According to him, on this day, at 6 o'clock in the afternoon, the world and time began.

Today, this unbelievably recent and precise date for the age of the universe is treated with derision and faint amusement, dismissed in popular accounts as the naive notion of a simplistic age. 'That Eastern Standard Time? Or Rocky Mountain Time?' asks Spencer Tracy, when a witness quotes Ussher's date in *Inherit the Wind*, the 1960 movie based on the famous 1925 'monkey trial' in Tennessee. Yet in Ussher's time, and for centuries after, the influence of his date was enormous. For nearly two hundred years, it was widely accepted as the true age of the world. It was printed in bibles, copied into various almanacs and spread by missionaries to the four corners of the world. For generations it formed the cornerstone of the Bible-centred view of the universe that dominated Western thought until the time of Darwin. And even then, it lingered on. As recently as the early 1900s publishers still printed bibles with '4004 BC' inscribed alongside the opening verse of Genesis; while echoes of its influence persist to this day. In a 1999 Gallup poll, 47 per cent of Americans said they believed that God had created human beings pretty much in their present form at one time within the last 10,000 years.

So how did people come to believe in such an improbable date? Certainly its widespread acceptance wasn't solely due to Ussher's standing as a scholar. Long before his time it had

10

been common knowledge that the world was created four to five thousand years before the birth of Christ. Shakespeare had even referred to it in *As You Like It*, when he gave Rosalind the line 'The poor world is almost six thousand years old.' All Ussher did was put a precise date to an idea that had been circulating in Europe and North Africa since the arrival of Christianity.

Even before Augustine's *City of God*, early Christians realised they had the means to put a date on the Creation. Scattered through the Old Testament are details of the lengths of people's lives which can be added together to indicate the number of years that have passed since the beginning. Genesis, for instance, tells the well-known story of how God created the world and its first inhabitants, Adam and Eve, but it also contains a lesser-known genealogy, a list of the male descendants of Adam.

> And Adam lived an hundred and thirty years, and begat
> a son in his own image; and he called his name Seth:
> And the days of Adam after he had begotten Seth were
> eight hundred years: and be begat sons and daughters:
> And all the days that Adam lived were nine hundred and
> thirty years: and he died.

The genealogy in Genesis runs through twenty-one generations of these long-lived descendants, all the way down to Abraham, and makes the chronologer's task easy by including, in each case, the age of the father when the first child was born. Simply adding these numbers together gives the span of time from the Creation to Abraham's birth.

By Ussher's time, the astonishingly long lives of Adam and his descendants (Methuselah, Adam's great-great-great-great-great-grandson, tops the list at 969 years) had come to be accepted as genuine. People regarded the early years of the world as a Golden Age in which the climate was healthier, food was plentiful and disease absent. As the first-century Jewish historian Josephus explained: 'their food was then fitter for the prolongation of life ... and besides, God afforded them a longer lifespan on account of their virtue'. Augustine explained that 'the earth then produced mightier men', and suggested that they had the stature of giants. What appears to have troubled him more than their longevity, though, was the fact that most of Adam's immediate descendants only became fathers after the age of a hundred. Was it credible, Augustine wondered, 'that the men of the primitive age abstained from sexual intercourse until the date at which it is recorded that they begat children'? Famous, himself, for saying: 'Give me chastity – but not yet!', Augustine concluded that they must have reached puberty much later than modern man.

While the dates given in this early part of the Bible are clear and straightforward, elsewhere they require more careful interpretation; nevertheless the genealogy continues right up to the time of the destruction of the temple in Jerusalem by the Persian king, Nebuchadnezzar. By searching through the Bible and adding together the relevant years, it was possible – or so many believed – to find the date for the beginning of the world.

Ussher realised this at an early age. Born in Dublin in January 1580, into a family of Irish gentry, young James was brought up as a Protestant, a rarity in a country that was largely Catholic. From birth he was surrounded by religion; his uncle, Henry Ussher, was the Archdeacon of Dublin, and the first book he learnt to read was the Bible, taught to him by two blind aunts who could recite much of it from memory. At the age of eight, his father sent him to a Latin school in Great Ship Street run by two Scotsmen as a cover for their secret activities as agents of James VI. Whatever their abilities as spies, they were excellent teachers, and within a few years Ussher could read and speak Latin fluently. This was a form-ative period. At the age of ten he was overwhelmed with a strong sense of religion (so much so, that a particularly beauti-ful sermon was said to have moved him to tears), and two years later he became fascinated with history. After reading a book on ancient empires, he carefully 'drew out the exact series of times when each Eminent Person lived', and three years later 'had made such a proficiency in chronology, that in Latine he drew up an exact Chronicle of the Bible, as farre as the book of Kings'.

And so began his quest to find the beginning of time: a quest that would occupy most of his life. For in drawing up his adolescent chronicle, he realised that determining the age of the world was no easy task. Although at first sight it appears simple – nothing more difficult than adding up all the dates in the Bible – in practice the unwary scholar soon becomes lost in a forest of conflicting dates and contradictory evidence. If Ussher ever thought that finding the age of the world would be a matter of a few months' or a few years' work, his illusions were soon shattered. By the time he entered Dublin's newly built university, Trinity College, in 1593 he probably already knew that, while many of the world's greatest minds had tackled the problem, none of them could agree on the answer.

The first person known to propose an actual date for the beginning of the world was the second-century bishop, Theophilus of Antioch (the city of Hatay in modern-day Turkey). At this time, the fledgling Christian Church was struggling to gain wider acceptance, but found its progress hampered by critics who questioned its validity. How could a religion barely a century old possibly be the true faith, they demanded, when Greek and Roman gods dated back as far as anyone could remember? It was a fair point, and one that Theophilus was familiar with. His pagan friend, Autolycus, had raised exactly that question. But the early Church fathers had already found an answer. Maintaining that the coming of Christ was the fulfilment of Old Testament prophecy, they adopted the Jewish books of the Old Testament as their own, and in one fell swoop Christianity acquired what was accepted to be the world's oldest and most detailed history. It was a master stroke.

Armed with this impressive lineage, Theophilus found it easy to answer his friend's challenge. Working through the Old Testament, he added up the dates of the patriarchs (the male descendants of Adam) and those of the judges and kings (who ruled Israel later), to show that: 'All the years from the creation of the world amount to a total of 5,698 years, and the odd months and days'. Therefore, Christianity 'is not recent, nor our tenets mythical and false, as some think; but very ancient and true'.

Once Theophilus had shown the way, others followed. Over the next fifteen hundred years, some of Europe's most distinguished theologians and philosophers attempted to find the date of Creation, but, despite their best endeavours, no two scholars arrived at the same result. Bede, in seventh-century England, named the year as 5199 BC; Martin Luther, in Germany, plumped for 4000 BC; while the seventeenth-century astronomer Johannes Kepler decided that 3992 BC was the most likely date. A later survey of works published in this period found 128 different dates for the Creation, ranging over three millennia. The youngest – 3761 BC – came from the chronology of the western Jews, while the oldest – 6904 BC – was derived by Alfonso, a thirteenth-century king of Castile. The lack of agreement was spectacular.

Given this evidence, it would appear that finding the birth date of the world was a hopeless task, inevitably doomed to failure. But Renaissance philosophers didn't see it that way, especially those in northern Europe. While the Renaissance in the south was characterised by an interest in painting and the arts – exemplified in the works of Leonardo da Vinci and Michelangelo – the emphasis in the north was on Christian scholarship of a penetrating kind. Scholars here regarded chronology as an important problem: an intellectual challenge worthy of the best and brightest minds of the day. Why else, they asked, would the holy writers fill the texts of Scripture with so many dates, if not to lay down a mystery for mankind to unravel? The wide range of dates, they believed, was the result of conflicting biblical texts and poor scholarship; by applying greater learning it was possible to find the true year of Creation.

Ussher himself was typically bullish: 'If any one, well seen in the knowledge, not onely of Sacred and exotick History, but of Astronomical Calculation, and the old Hebrew Kalender, shall apply himself to these studies, I judge it indeed difficult, but not impossible for such a one to attain, not only the number of years, but even, of days from the Creation of the World.'

On 20 December 1601, at the age of twenty-one, Ussher was ordained a priest. All of a sudden his interest in biblical chronology – previously just a youthful fancy – was sharpened by the need to argue the case for his Protestant faith.

In Ireland he faced an uphill struggle. Although Protestantism was the official faith in a country largely controlled by the English, most people were Catholic and stayed faithful to what they called the 'Old Religion'. They wanted nothing to do with the new-fangled religion with its 'Devil's Service' and its links with the English invader. 'None will come to my church at all,' complained the bishop of Cork. 'It is almost a bootless labour for any man to preach in the country but in Dublin for want of hearers.' But even in Dublin, the Counter-Reformation was in full swing. In the late 1590s, behind the backs of the authorities, Jesuit preachers had entered the city and begun celebrating the Catholic Mass in private homes. Their presence sparked a marked revival in the old faith – Ussher's own mother converted to Catholicism around this time – and the number of recusants (people who refused to attend the Protestant services) began to grow.

Ussher was luckier than other priests. Appointed preacher to the state, he gave the regular Sunday afternoon sermon in Dublin's Christ Church Cathedral, where he could at least rely on a congregation. Here in the city the Statute of Recusants was strictly enforced, under which anyone who failed to attend was fined one shilling. Despite this injunction, many leading citizens stayed away, accumulating fines of over £100, while those who did attend would 'walk round about like mill horses, chopping, changing, making merchandise, so that they in the quire cannot hear a word'. Only a skilled and determined preacher could cope with these conditions and, by all accounts, Ussher was both. He certainly held no sympathy for the Catholic point of view, and railed against it from the pulpit. The Church of Rome was the 'Babylon of the Apocalypse'; the Pope, the 'Antichrist'.

But he needed more than rhetoric to counter the charges levelled against his religion. 'Where was the Protestant Church before Martin Luther?', the Catholics asked, reminding him that it was less than ninety years since the German priest had nailed his 'Theses' to the door of the castle church in Wittenberg. How could such a new religion possibly be the true faith?

Even more damning was a book by Thomas Stapleton, an intellectual English Catholic. In *Fortress of the Faith* Stapleton claimed that the weight of historical evidence was on the side of the Roman Church. By selectively quoting the earliest Church leaders he asserted that it was Roman Catholic doctrine and not Protestantism that remained true to the teachings of the first Christians, and was therefore the one and only true faith.

Protestantism, he argued, was not merely new, its whole doctrine was false.

The accusation stung Ussher into action. The only way to counter these charges was to acquire a detailed knowledge of history and the Bible, a knowledge so extensive that he would be able to parry every thrust the Catholics aimed against him. To this end, he resolved to study. He would read the works of the early Church fathers, biblical commentaries, Church histories, everything he could lay his hands on. He would become an authority on the subject.

But he faced a formidable obstacle. Trinity College, at this time, was not the great seat of learning it later became; the university library held only forty books – hardly an adequate body of knowledge for a budding scholar.

By a stroke of good luck, however, help arrived from a most unlikely source. In 1602, having put down an Irish rebellion at the battle of Kinsale, the English army commemorated their victory by subscribing £1800 to buy books for the university. The college chose Ussher to travel to England to make the purchases, and the following year, together with his colleague, Luke Challoner – whose daughter he would later marry – Ussher set out for London.

It was a great opportunity. Anyone studying chronology needed access to the rare books and ancient manuscripts hidden in the libraries of the universities and the aristocracy. Spending a month each in London, Cambridge and Oxford, Ussher visited all the great libraries of England and familiarised himself with their collections. In London he gained access to the library of the wealthy collector Sir Robert Cotton, whose magnificent home on the banks of the Thames housed the finest private collection in the country – Francis Bacon and Ben Jonson were among the frequent visitors.

But not even the capacious shelves of Cotton's library could sate Ussher's appetite. He pursued books with an unparalleled vigour, obtaining catalogues for all the great collections. 'There was scarcely a choice book or manuscript in any of the libraries but was known to him,' wrote his chaplain. 'Nor was he conversant in the libraries of our own nation alone.' He employed scribes to make copies of the choicest books and manuscripts in the great libraries of Europe – the Imperial Library of Vienna, the Escorial in Spain, the Royal Library in Paris, and even the Vatican Library at the heart of the Roman Church.

Every three years Ussher repeated his journey to England, visiting the bookshops and libraries of London, Oxford and Cambridge, acquiring new works for Trinity College. Thanks to his efforts, in just ten years the forty volumes expanded to over four thousand. At the same time, he began building a formidable collection of his own. 'For books and learning he had a kind of laudable covetousness,' wrote a friend, 'and never thought a good book – either manuscript or printed – too dear.' By the end of his life he had amassed some 10,000 volumes: one of the largest private libraries in Europe.

Surrounded by his books, he threw himself into his studies. He read thoroughly the works of the earliest Christian teachers – a task that took seventeen years to complete; he plunged into the history of St Patrick and the first Irish Christians, aiming to show that their beliefs were identical to those of Protestants; and, finally, he immersed himself in the study of biblical chronology.

It was a long apprenticeship, but eventually it bore fruit. In response to Stapleton's *Fortress of the Faith* he wrote *An Answer to a Challenge Made by a Jesuit in Ireland*, an impressive 600-page tome that used quotations from the earliest Church leaders to establish that Protestantism was the true faith. And he followed this up with *The Religion of the Ancient Irish and Britons*, a book which showed that, in early times, the Christian religion in the British Isles was completely independent of

Rome. These works established his reputation as the foremost of the Anglican intellectual heavyweights, and his skills as a disputant were soon in demand.

A curious demonstration of his abilities is revealed by an unusual contest that took place in England between John Mordaunt and his wife, Elizabeth. Lord Mordaunt – a devout Catholic – was determined to convert his wife – a devout Protestant – to his faith. His wife, however, was equally determined to convert him to Protestantism. To settle the issue, they arranged an intellectual duel. To argue their respective cases, each side summoned the best advocate they could find. Lord Mordaunt chose an English Jesuit, known by the pseudonym of Beaumont; while his wife obtained the services of Ussher. At the end of 1625, at Mordaunt's seat in Northamptonshire, battle commenced.

Ussher spoke first. For three days he laid out his arguments, advocating his cause with such eloquence and conviction that, as each day passed, Beaumont appeared more and more nervous. On the morning of the fourth day, when it came to Beaumont's turn to speak, he was nowhere to be found; he had fled in panic. In his place he sent an apologetic letter saying he had 'forgotten all his arguments which he thought he knew as well as his Lord's prayer'. With no advocate to plead his case, Lord Mordaunt capitulated, and, to his wife's delight, converted to Protestantism. She never forgot Ussher's efforts on her behalf, and twenty years later would repay him in his hour of greatest need.

In January 1624, at the age of forty-four, Ussher was appointed archbishop of Armagh, the most senior position in the Church of Ireland. The post came with the benefice of a bishop's palace in Drogheda, thirty miles north of Dublin, and there he set to work on his most ambitious project – his universal history of the world.

Any history needed a starting point, and for Ussher, finding the date for the beginning of time was a natural progression from his battles with the Catholics. Unlike Theophilus and the early Christian chronologers, he had no intention of proving that his was the oldest religion. Such a task would have been futile; after all, both the Catholic and Reformed faiths were branches from the same tree. Instead, accuracy was his objective. He needed a date he could defend against the objections of the Jesuits, the most austerely intellectual of the Roman Catholic orders. If he could achieve this, he would establish the superiority of Protestant scholarship and the credibility of his faith.

The backbone of his history would be the Bible. But this presented him with his first dilemma. Which version to use? Different versions gave different dates. For example, the Greek-derived Septuagint Bible used by the Orthodox Church in eastern Europe gave dates that stretched back almost a thousand years earlier than those of the Hebrew-derived Bible used by the Catholic and Protestant Churches in the West. (The difference is mainly due to the fact that in the Greek text, Adam and his immediate descendants begat their offspring 100 years later than in the Hebrew.) For his chronology to have any credibility Ussher would have to show that the Bible he used was the true one.

Each version had its merits. The Septuagint, from the Latin *Septuaginta* meaning seventy, had, according to legend, been translated into Greek from the original Hebrew by seventy-two scribes – six from each tribe of Israel. This lent it not a little authenticity, but many Western scholars rejected it on the grounds that its chronology was too long. A strong tradition, stemming from an ancient Jewish text, held that the world would end in an apocalypse when it reached the age of 6,000 years; yet the dates in the Septuagint added up to at least 6,500 years. As the apocalypse clearly hadn't happened, they declared the Greek version unreliable.

The Bishop and the Book

The Hebrew text, on the other hand, came with an equally impressive pedigree. Back in the sixth century AD, Jewish scholars at the Talmudic schools in Chaldea and Palestine had collected manuscripts and oral testimony with the aim of reproducing as closely as possible the original text of the Old Testament. They assembled their evidence with meticulous care, checking every word and letter until they were satisfied that the finished script accurately conveyed the Word of God. To prevent errors creeping into future copies they counted the number of verses, words and letters in each section and noted which letter marked the middle of the text. Only then – four centuries after they started – did they declare the work complete.

For years, Western chronologers had favoured the Hebrew version, since both the Anglican and Catholic Churches used translations of its text. But in 1616, a remarkable discovery had called its veracity into question. In that year, Pietro della Valle, an Italian traveller passing through Damascus, had chanced upon an ancient and previously unknown version of the first five books of the Old Testament. It became known as the Samaritan Pentateuch, and word that its chronology was even shorter than that of the standard Hebrew text spread rapidly through the scholarly circles of Europe.

The discovery not only called into question the authenticity of the Bible, it raised the possibility that there might be other, even more ancient biblical texts waiting to be discovered – texts that might shed further insight into which was the true version.

Ussher was a thorough historian. One of his favourite sayings was a phrase of St Jerome's – 'Let those who care not for water from the purest source, drink from muddy streams.' He knew very well that the only true version of a text was the original itself; the later the copy, the more likely it was to be corrupted. Keen to track down further evidence before the

Jesuits got to it first, he took the radical step of employing an agent in the Middle East to seek out rare manuscripts.

Little is known of Thomas Davies, a Dublin merchant working at the vibrant trading centre of Aleppo in Syria, but his letters to Ussher paint a vivid picture of an agent eager to please. 'I should think myself happy', he wrote, after accepting the commission, 'that I were able to bring a little goat's hair or a few badger skins to the building of God's tabernacle.'

Ussher sent Davies instructions to track down and buy any ancient Bible texts in the local languages he could find. Davies, by return, warned him that: 'Such books are very rare, and esteemed as Jewels by the owners, tho they know not how to use them, neither will they part with them at dear rates, especially to strangers.' To overcome these difficulties, Davies dispatched messengers all over the Middle East: to Jerusalem, Damascus, Tripoli and even Mesopotamia, where 'there be found', he wrote, 'divers ancient books.' His efforts were soon rewarded and in August 1624 he wrote to Ussher with good news:

> The five books of Moses in the Samaritan character, I have found by a mere accident, with the rest of the Old Testament joined with them; but the mischief is, there wants two or three leaves of the beginning of Genesis, and as many in the Psalms, which not withstanding I purpose to send by this ship, lest I meet not with another.

This was a rare discovery and sheer good luck. With an attention to detail typical of his meticulous nature, Ussher immediately began learning Samaritan so that he could decipher the text himself. When it came to checking facts, it was said he 'would trust no man's eyes but his own'.

The Bishop and the Book

The following year, Davies found another version of the Bible:

> Amongst all the Chaldeans that lay in Mount Libanus, Tripoly, Sidon, and Jerusalem, there is but only one old copy of the Old Testament in their language extant, and that is in the custody of the Patriarch of the sect of the Maronites, who hath his residence in Mount Libanus, which he may not part with on any terms; only there is liberty given to take copies thereof.

On his own initiative, Davies dispatched a scribe to Mount Lebanon to make a copy of the Chaldean Old Testament. Back in Ireland, Ussher began learning another new language.

While Davies continued to scour the Middle East for new sources, Ussher began to tackle his second and more difficult problem – tying down the floating chronology of the Old Testament.

A floating chronology is a sequence of events whose dates are all known in relation to one another, yet the *time* when the sequence as a whole occurred is unknown. The Old Testament is a floating chronology: although it covers several thousand years of history, it doesn't say when the history occurred; it finishes before the New Testament begins, leaving a gap of indeterminate length between its conclusion and the birth of Christ. Anyone wanting to discover the age of the world had to find the length of this gap, and to do that required a formidable knowledge of the dates and events of ancient history. Unfortunately such details were scarce; the few contemporary accounts that survived were riddled with gaps, and Ussher, like other historians of the time, was constantly on the alert for fresh sources of information. In 1627 he received exciting news from England.

In January that year a ship had docked in London. On board

were over 200 pieces of ancient Greek marble acquired for the earl of Arundel, an avid art-collector. When the sculptures and tablets were unpacked at the earl's London home on the Strand, they caused a sensation. It was the first time ancient Greek inscriptions had been seen in England. Sir Robert Cotton, who was among those watching, became so excited that he rushed to the lawyer John Selden's house, woke him in the middle of the night and insisted he start deciphering the inscriptions at the crack of dawn.

Among the ancient stones laid out on the lawn of Arundel House was a piece of stone measuring just over a metre high by three-quarters of a metre wide. Found on the Greek island of Paros, it became known as the Parian marble. Carved on one side was a chronological inscription dating from the third century BC. A fascinated Selden later recalled that 'the decyphering of the marble of *Epochas* was the labor of a great many days . . . the characters being often entirely obliterated.' Eventually he uncovered a chronology of Greek history compiled by an author who declared that he had 'written up the dates from the beginning, derived from all kinds of records and general histories . . .' It listed the dates of a number of kings and archons (the chief magistrates of ancient Athens), and appeared to be just the missing link Ussher was looking for. The dates on the stone covered over a thousand years of Greek history, from 1582 BC to 264 BC, and the latter dates overlapped with the early part of the missing period between the Old and New Testaments.

When he heard the news, Ussher could barely contain his excitement. Writing to Selden, he declared that his own recent discoveries 'are nothing in comparison of the treasures which you have found of the Kings and Archons of Athens . . . You have made my teeth water at the mention thereof; and therefore, I pray you, satisfy my longing with what convenient speed you may.'

The Bishop and the Book

Unfortunately, the author of the Parian marble appears to have been a greater aficionado of plays and poetry competitions than of the reigns of kings. To Ussher's disappointment, much of the text was given over to unhelpful details such as the number of years 'Since comedies were carried in carts by the Icarians, Sufarion being the inventor, and the first prize proposed was a basket of figs, and a small vessel of wine.' Even worse, the earliest dates were for mythical events such as the trial of Mars for killing the son of Neptune. The Parian marble offered no instant solution to the problem of the floating chronology; instead, it merely drew attention to the enormity of the task ahead.

Ussher turned back to his books. The only hope of bridging the gap between the Old Testament and the year AD 1 was to piece together the chronologies of many different civilisations, and use them as stepping stones. Jumping from one history to another, then another, it might be possible to fill in the missing years. What made this task so difficult was that each nation had recorded its history using its own idiosyncratic system. The Jews, for instance, had used regnal years (the number of years from the start of a king's reign); the Greeks had counted in Olympiads (the four-year period between Olympic games); while the Romans had measured their early history in years elapsed since the foundation of Rome. To add to the confusion, in some countries the years weren't even 365 days long. The Arabs, for example, had used lunar years, which were based on the cycle of the moon and lasted just 354 days. And finally (to keep future historians on their toes), many nations began their years on different dates: so while the Roman year began with the winter solstice, the Attic Greeks had used the summer one, and the Egyptians had waited until the sun entered Aries. Faced with this confusion of dates, eras, epochs and seasons, Ussher turned to the pioneering work of a man who had trodden the same path fifty years earlier, the great Renaissance scholar, Joseph Justus Scaliger.

Scaliger, working first in Paris, then later at the University of Leiden in his native Holland, had devised a brilliantly simple solution to the chaos of the different calendar systems. He had invented a master calendar, the Julian Period, which began on a completely hypothetical day, 1 January 4713 BC, a day Scaliger was confident had never occurred. He had chosen it because all recorded events in history could be placed after it. What he had created, in effect, was the time equivalent of an exceedingly long tape measure alongside which the shorter yardsticks of all the different chronologies could be laid. It gave historians a powerful tool for determining when events occurred, and Ussher gratefully adopted it as the backbone for his chronology.

But the Julian Period wasn't the only technique Ussher borrowed from Scaliger. In an attempt to resolve the sometimes conflicting evidence of historical records, he unsheathed what would become the most powerful weapon in the chronologer's armory – astronomy.

The stars and planets are perfect timekeepers. (The year, after all, is defined as the time the Earth takes to orbit the Sun.) As the Earth follows its course, year after year, the position of the stars, moon and other planets appears to change in a regular predictable way; knowing this movement it is possible to calculate how the heavens would have looked at any moment in history.

Luckily for Ussher and other chronologers, the literature of ancient historians was peppered with references to astronomical events that could be dated. The summer and winter solstices, for example, were usually recorded within a few days of when they actually occurred, as were new and full moons. But the most important markers of time were eclipses.

Records of eclipses stretched back a long way – the Babylonians had recorded an eclipse as early as 763 BC. More important, they were sufficiently rare that if one was recorded as being visible from a certain place at some unknown time, it was almost always possible to pin down the exact year, day and hour it had occurred.

Although Ussher normally liked to check the facts for himself, it appears that astronomical calculations were beyond his capabilities. During his visits to Oxford he befriended John Bainbridge, the Savilian professor of astronomy (and a proponent of the new Copernican astronomy), who carried out the necessary calculations. 'I am in particular engaged in an expedite and resolute method of calculating eclipses, which I hope to accomplish to your Grace's content,' he wrote to Ussher in 1626.

One important date that Ussher decided on the strength of an eclipse was the birth date of Christ. When the seventh-century monk, Dionysius Exiguus, had devised the modern calendar, he had anchored it on the birth of Christ, which he believed had occurred in 1 BC. By the seventeenth century, however, mounting historical evidence suggested that Exiguus was wrong, and that Christ had been born earlier than this.

The evidence centred on the timing of King Herod's death. Matthew's gospel clearly stated that Jesus was born during the reign of King Herod; however, Herod was thought to have died before 1 BC. Although Scaliger maintained that Christ had been born in 1 BC, other historians plumped for 2 or 3 BC, while the astronomer Johannes Kepler put it at 4 BC or earlier.

Ussher agreed with Kepler. He decided the issue on the strength of an eclipse recorded by Josephus, in his account of the last days of King Herod:

As for the other Mathias who had stirred up the sedition, Herod had him burnt alive, together with his companions. And that very night there was an eclipse of the moon.

According to Josephus, this eclipse took place in the spring, shortly before Herod died – virtually marking the end of his reign, so Jesus must have been born before it. When Ussher checked the astronomical tables he found no lunar eclipses were visible from Judea in 3 BC. He also ruled out 2 BC, since according to Josephus the eclipse occurred in the spring, and the only lunar eclipse in 2 BC occurred in the summer. That left two options. Scaliger had favoured 1 BC, on the evidence of an eclipse that took place in January, but Ussher ruled this out, again because it was not the spring. Finally Kepler had chosen 4 BC on the strength of an eclipse that took place in March. When he checked his tables Ussher found that there had indeed been a partial lunar eclipse at 3 am on the morning of 13 March 4 BC. Later historians agreed, and although there is still some uncertainty, 4 BC is now accepted as the most likely year for Jesus's birth.

Around the time Ussher was grappling with Bainbridge's eclipse tables, an impoverished scribe at the rock-hewn monastery of Kenobin on the western slopes of Mount Lebanon was labouring to transcribe the Chaldean Old Testament. When he completed his work, he signed off with a poignant postscript:

Here ends this book by the help of our Lord Jesus Christ, in the year of Christ, 1627, in the month Thammuz [July], on the first day at the sixth hour, by the hands of a man sinful and vile, dust of the highways and dirt of the dunghill, the miserable Joseph, son of David, of the city beloved and blessed of Christ, Van of Mount Lebanon.

Even before he received the Chaldean manuscript, Ussher had three alternative texts for the Old Testament: his own Samaritan Pentateuch, the standard Hebrew text, and the Greek Septuagint. They each gave different dates, even for the simplest part of the chronology, the period between the Creation and the Flood. The Samaritan put this period at 1,307 years; the Hebrew 1,656 years; and the Greek 2,242 years.

To progress with his chronology he had to make a choice. The Greek version was easiest to rule out, on the grounds (mentioned earlier) that the apocalypse hadn't happened. Also, from his studies, Ussher had decided, according to the diarist John Evelyn, that it was 'full of errors'. That left the Samaritan and the Hebrew. The Samaritan script had its advocates, among them Sir Thomas Browne, who thought 'the Samaritans were no incompetent judges of times and the Chronologie thereof . . . and, as it seemeth, preserved the Text with far more integrity than the Jews'. However, Ussher's searches had failed to provide any evidence to back this up. If he chose the Samaritan he would be removing several centuries from world history on the evidence of a few manuscripts of uncertain parentage.

When Ussher received the Chaldean Old Testament he must have noticed with some degree of pleasure that its dates – for Genesis at least – matched those of the Hebrew Bible. It made his decision easy. He elected to base his chronology on the Hebrew text, a decision that, according to Evelyn, he regretted not making sooner: 'He told me how greate the losse of time was to study much the Eastern languages, that excepting *Hebrew*, there was little fruite to be gatherd of exceeding labour.'

For the next seventeen years Ussher devoted himself to writing and research, pursuing his studies quietly in his palace at Drogheda. A visitor at this time described him as 'a plain, familiar, courteous man, who spends the whole day in his study except meal-time'. He said prayers four times a day, and held chapel services before lunch and supper, but apart from that, his official duties were few. Outside the cloistered surroundings of his palace, however, Irish dissent was brewing. Events were about to take a turn for the worse, and, like a straw blowing in the wind, Ussher found himself at the age of sixty thrown into the most active and dramatic period of his life.

At dawn on 23 October 1641 the Irish Rebellion erupted throughout the land. By good luck, Ussher was in England with his wife and daughter at the time and so escaped the violence, but 'in a very few days the rebels had plundered his houses in the country, seized on his rents, quite ruined, or destroyed his tenements, killed, or drove away his numerous flocks, and herds of cattle, to a very great value; and in a word, had not left him anything in that Kingdom, which escaped their fury, but his Library, and some furniture in his house in Drogheda.'

But not even these were safe; the insurgents soon surrounded the town. 'We were besieged four moneths by those Irish Rebels,' wrote Nicholas Bernard, Ussher's chaplain, 'and when they made no question of devouring us [and] the library which I had the custody of . . . [they] talked much of the prize they should have of . . . burning it, and of me by the flame of the books, instead of faggots under me; but it pleased God in answer to our prayers, and fasting, wonderfully to deliver us, and it out of their hands.'

The rebellion left Ussher penniless. Deprived, not only of his home but also his income, he was forced to sell or pawn all the plate and jewels he had taken with him to England. He did, however, manage to retrieve his library, which was shipped to him in London in the summer of 1642.

The Bishop and the Book

While the violence in Ireland made it dangerous to return there, England was hardly any safer. That summer saw the first skirmishes of the English civil war, and autumn brought the first full battle. By winter the dividing line was clearly drawn: London became the stronghold of the parliamentary forces under Cromwell, while Charles I garrisoned his royalist army at Oxford. As the king's chaplain, Ussher's allegiance was naturally with the royalist cause, so, with parliamentary permission, he abandoned his library in London, packed two trunks and a single chest of books and travelled to Oxford to join the king. For two years, while the civil war raged all around, he lived at the university, preaching, counselling the king, and taking advantage of the excellent libraries to work on his history of the world.

With the chaos of the civil war all around him, Ussher none the less solved his knottiest problem: tying down the floating chronology of the Old Testament. The key lay in finding an event in the Old Testament that was also mentioned in the histories of pagan writers. He found the crucial link in the Second Book of Kings, chapter 25, verse 27:

And it came to pass in the seven and thirtieth year of the captivity of Jehoiachin king of Judah, in the twelfth month, on the seven and twentieth day of the month, that Evil-merodach king of Babylon in the year that he began to reign did lift up the head of Jehoiachin king of Judah out of prison.

Keen-eyed, Ussher spotted that the words '*in the year that he began to reign*' indicated that this was the year Evil-merodach's father, Nebuchadnezzar, had died. It was enough to bridge the gap. From a list of the kings of Babylon compiled by the second-century Greek astronomer Ptolemy, Ussher was able to link Nebuchadnezzar's reign to events in Greek history. It was then a simple matter to make the connection via Roman history to the modern Julian calendar. This put the date of Nebuchadnezzar's death as 562 BC.

Now he could calculate the age of the world. Beginning with Genesis and working through the Old Testament he added up the ages of the prophets and the reigns of the kings until he reached the death of Nebuchadnezzar 3,442 years later. Adding this to 562 BC gave him the date for Creation: 4004 BC.

It is a tribute to Ussher that his date for the death of Nebuchadnezzar is still accepted by historians today. It is curious, though, that he should have arrived at 4004 BC, for it gave exactly 4,000 years between the Creation and the birth of Christ – an appealingly round figure that happened to coincide with an ancient, but well known, Talmudic prophecy: 'The world is to exist 6,000 years. The first 2,000 are to be void; the next 2,000 years are the period of the Torah; and the following 2,000 years are the period of the Messiah.' Such mathematical symmetry fitted in with the commonly held view that God's universe was an ordered universe, and added greatly to the credibility of Ussher's date.

As well as naming the year of Creation, Ussher would also go on to fix its date and time, but Oxford in 1644 was becoming too dangerous a place to stay. The parliamentary army under Cromwell had gained the upper hand in the civil war, and at the insistence of friends Ussher headed west to the relative safety of Cardiff.

By the following summer the war was all but lost. Defeated at Naseby, the king also retreated to Cardiff, where Ussher was writing up the first part of his *Annals*. For a time the two men shared the same house. Ussher enjoyed 'his Majesty's excellent conversation', and the king in return treated him 'with wonted kindness, and favour'. But Charles was now on the run. On 3 August, Ussher preached to him for the last time. That evening he presented the king with copies of Psalms 100 and 101, and the following day bade him farewell. As Charles marched out of the city, he took with him the entire garrison and all the ammunition.

With Cardiff unprotected and the parliamentary army advancing, there was little Ussher could do but flee. Accepting an invitation from Lady Stradling to stay at her castle in St Donats, twenty miles away, he headed west, accompanied by his daughter, his chaplain, and a chest containing his books, papers, and the half-completed *Annals*. It was a disastrous decision.

The countryside around Cardiff at this time was swarming with bands of armed Welshmen claiming to support the king, yet refusing to join his army. As Ussher and his companions approached St Donats they rode into an ambush. The local militia 'immediately fell into plundering,' recalled his chaplain, 'breaking open my Lord Primate's chest of books, and other things which he then had with him, ransacking all his manuscripts and papers, many of them in his own hand writing; which were quickly dispersed among a thousand hands; and not content with this they pulled the Lord Primate, and his daughter, and other Ladies from their horses.'

Fortunately, at that moment the officers of the rabble happened to ride up. As members of the local gentry, they were ashamed at the ill-treatment meted out by their men, and immediately ordered them to return the horses. But the real damage had already been done; Ussher's papers, his life's work, were blown to the wind, scattered too widely to be retrieved. 'I must confess that I never saw him so much troubled in my life,' his chaplain recalled, 'he seemed not more sensibly concerned for all his losses in Ireland, than for this.'

Comforted by his daughter, Ussher was helped to a nearby house, scarcely able to believe that such a disaster could be the will of God. 'He has thought fit to take from me at once,

33

all that I have been gathering together, above these twenty years, and which I intend to publish for the advancement of learning, and the good of the Church,' he wailed.

The next morning several members of the local gentry visited him and promised to do their best to recover any papers that had not been burnt or torn to shreds. Advertising the loss during church services, they asked 'that all that had any such books, or papers, should bring them to their Ministers, or landlords'. Amazingly, little by little, the documents began to be handed in, 'so that in the space of two or three months there were brought in to him, by parcels, all his books and papers, so fully, that being put altogether, we found not many wanting'.

But Ussher was now trapped. To the north and east the parliamentary forces were closing in. In desperation, he chartered a ship, hoping to cross the sea to France, but before he could set sail, a parliamentary fleet drew into Cardiff Bay, cutting off all hope of escape. Completely surrounded, and at the mercy of the parliamentary army, he had resigned himself to capture, when, out of the blue, he received an unexpected offer of help.

Lady Mordaunt – now the countess of Peterborough – whose husband Ussher had converted to Protestantism all those years ago, hadn't forgotten his lengthy advocacy on her behalf. She was now in a position to repay the favour. At the start of the civil war, her husband had sided with the parliamentary cause, and although he had died a few months earlier his widow continued to hold influence in high places. When she heard that Ussher was stranded in Wales, she sent a message inviting him to stay at her Charing Cross home, and arranged for his safe passage to London.

The capital had changed dramatically in the four years since Ussher's last visit. Many of the owners of the great houses had fled the city for the safety of the Continent. Arundel House, where Cotton and Selden had rushed to examine the inscriptions on the marbles, had been abandoned, and was now occupied by a garrison of Roundhead soldiers. The marbles still remained – though around this time the Parian chronicle was badly damaged when part of it was used as a hearthstone. Elsewhere in the city, Ussher's library was still intact; a handful of friends had prevented the parliamentary forces from breaking it up. Reunited with his books, and under the protection of the countess of Peterborough, Ussher collected together his manuscripts and launched into the final stage of his history.

The years had taken their toll. Half blind from a lifetime of reading, he spent each day shuffling around the house, following the sun from room to room, in order to glean enough light to write. Having fixed the year of Creation, he now went on to determine its precise date and time.

Precisely which time of year Creation had occurred had long been a subject for debate. For centuries there had been a common belief that God had created the universe at the exact moment when the sun was at one of its four cardinal points: either the winter or summer solstice, or the spring or autumn equinox. This made sense; it gave the universe an astronomical symmetry, that agreed perfectly with the view that God's universe would display mathematical harmony.

But which one of the four? While a few chronologers opted for the summer solstice, the equinoxes were by far the most popular choice. Among the scholars of the early Christian Church, spring received the most votes – presumably because it was the season of growth and renewal. Geoffrey Chaucer reflected this view in 'The Nun's Priest's Tale':

Whan that the month in which the world began
That highte March, when God first maked man.

By the seventeenth century, however, the balance of opinion had swung to autumn. Autumn marked the start of the Jewish year, but there was another, more logical explanation. It was also the time of harvest, and biblical scholars couldn't fail to notice that when Adam and Eve arrived in the garden of Eden, the fruit was ripe and ready for picking.

Having assumed that the world began in the autumn, Ussher took it for granted that the first complete day of the world would be the first day of the week – a Sunday. Having made all these assumptions, and knowing the year to be 4004 BC, calculating the date was straightforward: 'I have observed that the Sunday, which in the year [4004 BC] aforesaid, came nearest the Autumnal Aequinox, by Astronomical Tables, happened upon the 23 day of the Julian October.' While this gave Ussher his first *whole* day, pedantic to the last, he argued that *time* began a little earlier. The explanation lies in his interpretation of the first verses of Genesis:

> In the beginning God created the heaven and the earth.
> And the earth was without form, and void; and darkness
> was upon the face of the deep.

It was clear from this, argued Ussher, that the world was dark when God created it; light only came later. Just how much later was revealed a few lines further on:

> And the evening and the morning were the first day.

He took this sentence to mean that the first day *began* with evening. According to Ussher, 'from the evening preceding, that first day of the Julian year, both the first day of the Creation, and the first motion of time are to be deduced.' In other words, as he made explicit in his introduction, time began at 6 pm on the evening of Saturday 22 October 4004 BC. In later years most commentators ignored this subtle refinement, and so, in many books, 23 October 4004 BC has come to be regarded as 'Ussher's date'.

At just after 2 o'clock on the afternoon of 30 January 1649 Charles I walked out onto the scaffold in front of the Banqueting Hall in Whitehall. At Lady Peterborough's house the spectators on the roof finally persuaded Ussher to join them, 'as much out of desire to see his Majesty once again', his servant recalled, 'as also curiosity, since he could scarce believe what they told him, unless he saw it'. Ussher's history had depended on the lives of kings: biblical kings, Greek kings, Babylonian kings. It was the ages of their lives that had enabled him to construct a chronology of the world. Now, his own king was about to die. As Charles I made his final speech, Ussher's eyes filled with tears. He lifted his hands to heaven and prayed earnestly. Far below, the king finished speaking, removed his cloak and doublet, and lowered his head onto the block. This last sight of his old friend was too much for Ussher to bear. As one of the executioners lifted the hair from the back of the king's neck, he turned pale and collapsed to the roof in a faint.

Just over a year later, in the summer of 1650, the first part of the *Annals* – including the date of Creation – went on sale at the sign of the Ship, one of the many booksellers plying a trade in St Paul's churchyard in London. The book immediately confirmed Ussher's reputation as the foremost biblical scholar of his day. Four years later he completed the second part, bringing his history of the world up to AD 70. He contemplated a third, but by this time had become too frail to continue. In January 1656 he wrote in his almanac: 'Now aged seventy-five years. My years are full', and underneath, in large letters, added 'Resignation'. The following month he travelled to the countess of Peterborough's country house in Reigate where, on 21 March 1656, he died.

When Cromwell heard of Ussher's death he ordered a state funeral with full honours in Westminster Abbey, an indication of the esteem in which Ussher was held. Although a great honour, the expense of the service came to considerably more than the £200 Cromwell contributed, and it was left to Ussher's family – who had hoped for a quiet ceremony in Reigate – to pay the difference.

Cromwell also refused to allow Ussher's magnificent collection of books to be sold without his consent. He rejected an offer from the king of Denmark in favour of one from the English army in Ireland, led by his son, Henry. Ironically, the army that caused so much bloodshed and destruction as it repossessed Ireland after the rebellion, now contributed to the country's heritage. With the sum of £2,200 deducted from the officers' pay, the army bought the complete collection of books, crated it up, and shipped it to Ireland. After a brief delay, it arrived in Dublin's Trinity College, where it remains to this day.

Despite the accolades showered on his work by his contemporaries, Ussher's date for the Creation of the World would have sunk into perpetual obscurity – like the dates of the hundred or so chronologers before him – if it hadn't been for a London bookseller named Thomas Guy. In about 1675, Guy, an enterprising businessman, contracted with the University of Oxford for the right to print bibles under their licence. As a marketing ploy he printed Ussher's chronology in the margin, thereby enabling readers to see at a glance when all the events in the Old and New Testaments had taken place. The new bibles were an immediate success; possibly helped by the inclusion of dramatic illustrations of Bible stories, including – in true tabloid style – engravings of bare-breasted women. Sales boomed, earning Guy a small fortune which he subsequently invested with great success in the infamous South Sea Company. By the time he died he had made enough money to endow the famous London hospital that still bears his name.

But that was only the beginning. In 1701 Ussher's chronology received the blessing of the Church of England itself when William Lloyd the bishop of Worcester authorised its use in an official version of the Bible. Once inside these holy pages, Ussher's dates practically acquired the authority of the word of God. They quickly became the 'Received Chronology', adopted by nearly all the Reformed Churches, and within a few generations had become such an integral and familiar part of the Bible that most people no longer remembered where they had come from. Consequently, they continued to be printed in the margins of bibles right into the twentieth century. The year 4004 BC cast a long shadow; it reinforced the widespread belief in a young universe and, for the next two hundred years, continued to influence the way natural philosophers viewed the world.

3. Doubters

Supported by the authority of scholarship, the approval of the Church and the lack of any alternative explanations, Ussher's pronouncement appeared virtually unshakeable. As long as the foundation on which it was built – the authority of the Bible – remained uncontested, there was no reason to doubt its validity. However, even as the ink was drying on his manuscript, a small number of free-thinkers were beginning to question the Holy Book, and the chronology it contained.

During the previous century Europeans' horizons had broadened. The great voyages of exploration had opened up the world to travellers who returned from far-flung countries with stories of how *their* histories stretched back as far as, if not further than, the sacred history of the Bible. While these tales were frequently exaggerated and invariably lacked documentary evidence, they nevertheless raised the possibility that the world was older than Europeans imagined.

On their own, these stories would have posed little threat to the Bible chronology, supported as it was by the greatest scholars of the day. Combined with cracks in the biblical account, however, they inspired the first serious assault on the Hebrew timescale.

In the vanguard of the assault was Isaac La Peyrère, a French Calvinist and lawyer, who came from a noble Bordeaux family. In 1641 La Peyrère sought the permission of

Cardinal Richelieu, the First Minister of France, to publish a manuscript he had recently completed. It was a rash move. His treatise proposed the heretical notion that people had existed before Adam. Whatever Cardinal Richelieu thought when he read La Peyrère's manuscript, he was no fool. The suggestion that other races had inhabited the earth before Adam and Eve called into question the whole biblical account of Creation, and the publication of such an idea could only damage the Church. Anticipating the controversy the book would arouse, he promptly banned it. It would be fourteen years before La Peyrère again plucked up the courage to reveal his treatise to the world.

According to La Peyrère's own account, the inspiration for his controversial idea came from the Bible itself – from discrepancies in its narrative he considered so glaring he had noticed them while still a child. For instance, how did Adam's son Cain find a wife unless there were already women in the world? And why did God have to mark Cain, 'lest any finding him should kill him', if the only people in the world at the time were his parents and siblings, who would surely know who he was?

By 1640 La Peyrère had developed his ideas into a fully-fledged theory. While he acknowledged that Adam had been created by God, he argued that he was not the father of all mankind, as the Bible claimed, merely the founder of the Jewish race. All the other races of the world, he believed, had been created at some earlier time. Exactly how much earlier, La Peyrère didn't say, but he implied that it had to be many hundreds of thousands of years before the creation of Adam.

For evidence, he drew on ancient accounts of Chaldean and Egyptian history, as well as more recent reports arriving from Peru, Mexico and China. Where other scholars had dismissed any dates older than those in the Bible as fictitious, La Peyrère built his case on the earliest dates he could find. He championed the reported 50,000-year antiquity of the Egyptians, and the 470,000 years that the Chaldeans claimed had passed 'since the time they had begun to observe the stars'. More scholarly academics dismissed such figures as gross exaggerations. 'Incredible was this absurd vanity of the Egyptians, who, to make themselves the first of the creation, lied so many thousand yeares.' But La Peyrère maintained they told the truth.

On its own, this tampering with sacred history was enough to arouse the wrath of the Catholic Church, but then La Peyrère went even further. He argued that the universal Flood – which according to Genesis destroyed all the people of the world except Noah and his family – had been in reality a regional flood that only destroyed the Hebrew people. The events described in Genesis, La Peyrère argued, were only local events, and by implication the biblical chronology that Ussher and others had put their faith in was nothing more than local history.

As for the beginning of time, that was so far in the past as to be immeasurable. 'I dare boldly affirm we know not that beginning,' he wrote. 'I know there is a settled number of stars in heaven, there is a determinate number of the grains of sand in the Sea shore: But I think to make up a sum of all those stars, all those grains, all those ages which have bin from the beginning, is without the compasse of all Arithmetick and humane account.'

Without the support of a powerful ally, publication of such radical views was unthinkable. In 1655, however, La Peyrère found an influential patron.

The previous year, Queen Christina had dramatically abdicated her throne in Sweden and moved to Antwerp. The unexpected arrival of this intelligent and witty woman caused a sensation in the city. La Peyrère's master, the prince of Condé, was so won over that he determined to win her hand in marriage. He installed La Peyrère in a house next to Christina's, and arranged for him to carry secret messages back and forth. Christina was charmed. Not by the prospect of marriage to Condé, but by La Peyrère's thesis, which he had seized the opportunity of reading to her during their meetings. A firm believer in free speech and a great supporter of the arts, Christina urged him to go to Holland, where it was easier to publish such radical ideas. He took her advice, and in 1655 *Men Before Adam* was finally printed in an edition almost certainly financed by Christina. Fearing recriminations, however, both La Peyrère and the printer omitted their names.

The book was a sensation. As rumours circulated of its scandalous content, people scrambled to get copies: 'it flies in an instant through the Christian world,' wrote one commentator, 'and is not only sold to the highest bidder, but is fought for among the buyers.' The reaction among scholars was sheer outrage. One writer described it as a 'pestiferous work', while another thought it 'should have remained buried in an eternal night'. Over the next fifty years at least thirty-eight different authors published refutations of its argument.

Although La Peyrère published his book anonymously, he had spoken of his hypothesis in public so often that he was instantly recognised as the author. Trouble was inevitable. On Christmas Day 1655, the bishop of Namur, the Belgian town where La Peyrère lived, publicly condemned *Men Before Adam* in all the town's churches, and banned the 'reading, holding, and selling of the book, as it contains heretical, erroneous, and rash affirmations'. But worse was to come.

On a cold February day the following year, a group of thirty armed men burst into La Peyrère's house and arrested him. In jail they repeatedly questioned him about his views and demanded he repent. For four months he held out, refusing to retract a word, until eventually his interrogators offered a deal: if he converted to Catholicism and apologised to the

Pope he would be forgiven. In June 1656 La Peyrère accepted the offer, and the following winter travelled to Rome to make his apology in person. He escaped lightly. When he presented himself at the Vatican, Pope Alexander VII greeted him with a mocking smile. 'Let us embrace this man', he joked, 'who is before Adam.' And the General of the Jesuits, who was also in attendance, revealed to La Peyrère that both he and the Pope had laughed wholeheartedly when they read his book.

Despite this brush with the authorities, La Peyrère clung stubbornly to his belief. He carefully phrased his abjuration in Rome so that he never actually *said* his theory was wrong, and when, in 1665, he retired to a seminary on the outskirts of Paris, he continued to collect evidence to support his case. It's easy to see why the Pope didn't take him seriously. In the encyclopaedic tradition of a bygone age, he gathered together all the arguments he could find, and presented them with no regard as to whether they were rational or not. His 'evidence' therefore included the curious notion that Adam had suffered from gout – a hereditary disease – and the observation that the Bible began with the second letter of the Hebrew alphabet, Beth, and not the first letter, Aleph. This apparently implied there had once been an earlier book beginning with the letter Aleph which was now lost, and which doubtless told of men before Adam.

Faced with such arguments, the Catholic Church could laugh off La Peyrère's theory with confidence. His evidence was held to be unreliable, he commanded no band of supporters, and his suggestion that Chaldean history stretched back 470,000 years was so far removed from the accepted 6,000-year history of the world that it was easily dismissed as a myth.

But while La Peyrère's hypothesis was publicly rejected by nearly everyone, it provoked a debate in Europe that continued for over fifty years. *Men Before Adam* opened people's minds to the possibility that the world might be older than they thought. And while the Vatican could dismiss it as fantastic in 1656, time would endorse La Peyrère. Within a few years, new evidence emerged that supported his hypothesis, and once again set Europe ablaze with argument.

Martino Martini.

In 1642 the Jesuit missionary Father Martino Martini arrived in China. For ten years he travelled the country, mapping its towns and villages, learning its language, and preaching the Catholic faith. Like most Jesuits, Martini was eager to understand the land in which he worked, and he went out of his way to learn about its past. He spoke to teachers, collected Chinese manuscripts and consulted with the most authoritative scholars he could find. However, the more he learnt, the more he realised the scale of a problem that had dogged the Jesuits' attempts to teach Christianity since they had first arrived in the country seventy years earlier.

The problem was one of chronology. When the first Jesuit

The family tree of Noah.

missionaries had arrived in China, the Chinese had greeted their account of world history – drawn from the Hebrew Bible – with disbelief. The cause of this dissent was the story and date of the biblical Flood, the fearsome deluge in which, according to Genesis, all the people of the world had drowned except Noah and his family. According to the missionaries, this punishment from God had occurred in or around the year 2300 BC. The Chinese, however, dismissed this as impossible. Their own history stretched back hundreds of years before this date, and made no mention of a worldwide flood, let alone a man called Noah. Either the Flood hadn't reached China, or it had occurred much earlier than the Jesuits said.

The Jesuits had soon realised that any attempt to teach Christianity using the chronology of the Hebrew text was doomed to failure. In 1637 they had astutely obtained the

Pope's permission to use instead the longer chronology of the Septuagint Bible, in a version which placed Creation at 5199 BC and the Flood at around 2950 BC. Although this 'solution' had initially horrified European theologians, they had assumed that it was only a temporary measure. As soon as the Chinese histories were properly examined, they believed, they would be found, like those of the Egyptians and other nations, to be false.

In 1651 Martini was ordered to undertake a diplomatic mission to the Vatican. As he boarded the ship in Zhangzhou for the return journey to Rome, he took with him over fifty original Chinese texts, including histories that he believed would reveal the truth of Chinese chronology. At this time no accurate account of Chinese history had yet been published in Europe, and the scale of the discrepancy between Chinese and European chronologies was unknown. Determined to correct this ignorance, Martini resolved to write a history of China himself.

As he read through the texts, however, Martini realised that the Chinese chronology posed a serious threat to the authority of the Bible. The dates of the Chinese Imperial dynasty indicated that the first emperor, Fu Hsi, had begun his rule in 2952 BC, some 600 years before the Hebrew date for the Flood. They also revealed a continuous line of descent ever since, with no gaps in the chronology. Martini was impressed. The Chinese records appeared more detailed and reliable than those of any other culture, including the Jews. They contained no bizarre myths or unbelievable legends, the lengths of the emperors' reigns were clearly documented, and even the earliest accounts had been written by contemporary authors. With the intellectual honesty typical of his Jesuit training, he concluded the records were genuine, and resolved to break the news to Europe.

It took Martini three and a half years to reach Rome. In

the Philippines he was captured by the Dutch, who held him prisoner for over a year. Then, on his release, he took passage on a ship returning to Holland, but which was forced to sail around the northern tip of Scotland to avoid the Anglo-Dutch war in the English Channel. Finally, after his vessel was caught in a storm and blown onto the coast of Norway, he was left to make his way across Europe by land. When he finally arrived in Rome, in the autumn of 1654, he was kept busy at the Vatican with important diplomatic duties, so it wasn't until 1658 that his history was finally published.

Like the response to La Peyrère's *Men Before Adam*, published three years earlier, the reaction to the news that China's history pre-dated the Flood was overwhelmingly hostile. Most academics immediately assumed that the Chinese had exaggerated their history out of national pride, and that the records Martini had consulted were not authentic histories, but recent fabrications written to establish the greatness of their nation. How else, they asked, could they have survived the ruling of a third-century emperor who – according to China's own history – had ordered all the books in the country to be burnt?

But what astonished Europeans almost as much as this detailed chronology was the revelation that China was a great civilisation. For years Europeans had regarded their own continent as the centre of the civilised world, yet here were another people possessing an organised civic structure, strong moral codes, and high culture: attributes Europeans liked to associate with their *Christian* civilisation. Perhaps, some scholars suggested, both cultures shared a common root.

One of the first to propose this idea was the German naturalist and polymath, Athanasius Kircher. He argued that the Chinese were the descendants of Ham, one of Noah's sons, who had left the Middle East and migrated to China. Taking this idea one step further, the English architect (and pupil of Inigo Jones) John Webb declared that Chinese was the original language of mankind that had been spoken by Adam and Eve in the Garden of Eden. The Chinese written characters, he noted, looked similar to those in the Hebrew alphabet; the Chinese wrote their words from the top down, as was common in ancient hieroglyphics; and finally, the Chinese couldn't pronounce the letter 'R'. Pronunciation of this particular letter, Webb explained, had to be drummed into European children, therefore the natural primitive state was not to be able to pronounce it.

Martino Martini didn't stay in Europe long enough to see his book published, let alone join in the debate. After a brief stay in Rome, he left Europe in 1657 to return to his missionary work in China. Despite an attack by pirates, numerous storms and a disease that killed twelve of the seventeen missionaries accompanying him, he eventually reached his mission in Hangzhou. Three years later he was taken ill with severe constipation. Ignoring the advice of a local Chinese herbalist, he prescribed himself a large measure of the cathartic form of rhubarb, overdosed, and died. According to a local legend, his body was preserved at the mission, and, for years after, was brought out on special occasions and seated in a place of honour.

The controversy Martini left behind continued to simmer for nearly a century. In response to the claims of European scholars that the Chinese had fabricated their history, the Jesuits in China scanned the ancient texts for records of eclipses. By 1729 they had found 26 solar eclipses that 'according to calculation fell on the exact year, month, and day indicated by the Chinese authors'. The earliest of these dated back to 2155 BC, only 250 or so years after the Hebrew date for the Flood. For the Peking Jesuits, this proved that the Chinese chronology was correct, but it wasn't enough to convince their European colleagues.

Indeed, Europe seemed quite immune to any change:

neither La Peyrère's theory nor Chinese chronology even dented the popular belief in Genesis, or the widely held conviction that the world was only 6,000 years old. In China, however, as it became evident that the dates in the Hebrew-based Bible had to be wrong, the Peking Jesuits grew increasingly frustrated and baffled by their European colleagues' reluctance to abandon the Hebrew chronology. In 1740, Father Perrenin, a Jesuit missionary working in Peking, wrote despairingly:

> I dare to hope that the Hebraicising gentlemen will permit us to lengthen the duration of the world by a little . . . It is much easier to persuade the astronomers than the chronologists . . . There is not much hope that they will be touched by astronomical proofs, or historical proofs, or the proofs of physics. The scholars . . . have published large volumes on chronology and each one of them has done his best to prove himself right. They cannot agree among themselves, and if you dare to interfere in their disputes with arguments about far-off countries, they all jump on you and not a one of them will concede you a month of time or an inch of terrain to carry out your evolutions.

By the time Father Perrenin wrote these words, however, the rules of the game had begun to change. For centuries, scholars researching the events of the early world had based their inquiries exclusively on the study of ancient texts. Words – either in the Bible, the writings of early historians, or the inscriptions on classical monuments – had been the keys to the past. The seventeenth century, however, saw a crucial intellectual shift: a move away from this dependence on textual authority towards a new form of inquiry based on scientific principles – natural philosophy. '*Nullius in Verba*' – 'Take Nobody's Word for it' – proclaimed the motto of Britain's

Royal Society, established in the 1660s for the pursuit of this new philosophy. Words handed down from ancient authorities were no longer to be trusted; philosophers were expected to draw their own conclusions from observation of the natural world. Throughout Europe, the rallying cry was the same; rocks, not books, were believed to hold the secrets to the past. It would no longer be historians, but natural philosophers, who proclaimed the age of the world.

4. Change and Decay

It could be said that there were two chronologies in the Bible: one describing the generations of mankind – the almost six thousand years of human existence – the other, much shorter, describing the six days it took to create the world. Up to the beginning of the seventeenth century, scholars had devoted almost all their attention to the later chronology, the span of man's history, but from now on their focus turned increasingly to the earlier period, the time it took to build the world. The more they probed this question, the older the world became.

The driving force behind their inquiries was the idea of *change*. In the previous hundred years there had been a crucial shift in the way people interpreted the biblical story of the Creation. Earlier, in the Renaissance, most scholars imagined that God had created the world fully formed; it came complete with mountains, streams, valleys – everything that Renaissance scholars could see around them. By the 1630s, however, that view had largely vanished. Scholars had increasingly come to believe that the earth had *aged* since its creation. A perfect God, they believed, would have created a perfect world, which meant a perfect sphere. The idea was popularised by the Spanish humanist Antonio de Torquemada, who in 1614 explained that 'the whole world before the time of the flood was plaine and levell, without any hill or valley at all.' As evidence, he pointed out that the Bible made no mention of mountains until after the Flood.

At this time, therefore, most scholars' view of the world and its history went something like this. That the earth had been created by God was indisputable. It was also probable that, when first created, it was a perfect sphere, smooth and uniform all over, as unblemished as the skin of a new-born child. The early days of this young planet – in the time before the Flood – had been a Golden Age, an era of balmy climate, bounteous fruits and great longevity; an age when God had blessed the world. But by the seventeenth century the earth was in its dotage. 'The World it self,' wrote Sir Thomas Browne, 'seems in the wane, and we have no such comfortable prognosticks of latter times, since a greater part of Time is spun than is to come.' This belief, originating in the prophecy that the world would only last six thousand years, was confirmed by nature. Mountains and valleys appeared as unsightly wrinkles, wrought by the earth's old age; volcanoes and deserts were boils and blemishes on what had once been a perfect landscape.

The key concept of this idea was that the earth was not a static object but was dynamic and had altered over time. This idea of change, and in particular the determination of the rate of change, would play a key part in the new chronology. But before these processes could be examined, Western philosophy had to break free of the literal interpretation of the Bible. Only by breaking with the Bible could the age of the world be pushed back in time, but such a break called for a radical new philosophy: a completely new way of viewing the world.

The foundation of this new philosophy was laid down twenty years before Ussher published his *Annals*. Its architect was a short, nervous Frenchman – René Descartes.

In 1619, at the age of twenty-three, Descartes had set his sights on becoming a philosopher. For six years he had travelled through Europe, responding to anyone who asked what he was doing by explaining he was studying the 'book of the world'. Quietly, though, he was developing ideas that would revolutionise the way mankind viewed the universe – ideas that broke away from the prevailing dependence on the Bible, and insisted instead on reason.

In 1625, Descartes arrived in Paris. Clever and sociable, he was quickly drawn into the intellectual life of the city. He gambled, attended court, visited the theatre, but above all he debated philosophy with his friends. The French capital at this time was a centre of radical thought: libertine views flourished as free-thinkers began to challenge the accepted wisdom of the establishment. Descartes was in his element, taking on all comers in argument and debate, but he was careful not to air his deepest thoughts too freely; the libertines trod a fine line between being tolerated by the Church, and being condemned.

The Catholic Church wielded great power. It was not beyond punishing heresy with the severest sentence. Although most philosophers were believers, those that persisted in challenging the Church's authority could expect little sympathy. In 1600 Giordano Bruno – who had argued that the universe was infinite – was burnt to death in Rome for denying the divinity of Christ; in 1619 the Inquisition at Toulouse had executed Giulio Vanini – a doctor who had proposed a natural explanation of miracles – by cutting out his tongue, strangling and burning him. Shortly before Descartes arrived in Paris, the Church had accused an acquaintance of his, Théophile de Viau, of writing poetry that mocked religion. De Viau had fled the city, but in his absence he was sentenced to death, and his effigy burnt.

This was clearly no place to develop controversial ideas, and in 1628 Descartes left Paris for the Protestant Netherlands, where greater religious freedom allowed him to pursue his thoughts unmolested. 'Here I sleep ten hours every night,' he wrote, 'without being disturbed by any care.' In this quiet retreat, away from the hubbub of Parisian life, he determined to rethink philosophy from scratch: to build up a view of the world and how it had formed through reason alone. Unlike *natural* philosophers, for whom experiment was essential, Descartes believed he could explain the universe just by thinking. '[T]here is nothing so remote,' he wrote, 'that it cannot be reached, nothing so hidden that it cannot be found.' Even an event as distant as the beginning of the universe would, he believed, reveal its secrets to his powers of deduction.

In 1630, he began writing a book called *The World*, in which he aimed to show how the initial chaos at the time of Creation, through a series of rational processes, could create the earth, planets, stars and everything else in the universe. He set himself a deadline of completing it within three years. The challenge was clearly daunting, as he revealed in a letter to his friend, Marin Mersenne, a priest and mathematician living in Paris, and the hub of a correspondence network of natural philosophers. 'I am now ordering the chaos to produce light,' wrote Descartes. 'This is one of the highest and most difficult tasks that I could face because it involves virtually the whole of physics. I have to think of a thousand different things at the same time in order to find a way of stating the truth without shocking someone or offending received opinions. I want to take a month or two to think of nothing else.'

While he worked on *The World*, Descartes was secretive to the point of paranoia. He moved house frequently, refused to tell his friends where he was staying, and sometimes wrote false return addresses on his letters to avoid unwelcome visits. The only person he regularly confided in during this time was Mersenne. In a letter dated July 1633 progress seems to have been going well, for he told Mersenne that *The World* was 'just about finished', and promised to send him a copy by the end of the year.

That autumn, however, Descartes heard news that threw him into a panic. The Italian astronomer Galileo Galilei had been arrested in Rome. His crime – publishing a book proclaiming that the Earth travelled around the Sun. Although this wasn't a new idea (the Polish astronomer Nicolai Copernicus had suggested it over a century earlier), it still contradicted Church doctrine which maintained that the Earth was stationary at the centre of the universe. Galileo was forced to recant his views and sentenced to house arrest.

'I was so astonished,' Descartes wrote to Mersenne, 'that I have almost resolved to burn all my papers, or at least not to show them to anyone.' His alarm was understandable: he had incorporated Copernicus's theory into *The World*. 'I confess that if this is false, then all the principles of my philosophy are false also,' he wrote. 'And because I would not want for anything in the world to be the author of a work where there was the slightest word of which the Church might disapprove, I would rather suppress it altogether than have it appear incomplete – "crippled", as it were.'

By February 1634 he had made his decision: 'I have decided to suppress my treatise entirely,' he told Mersenne, 'and thus lose almost all of my labour during the past four years in order to render entire obedience to the Church.' While *The World* would never be published in his lifetime, Descartes was too proud of his theory to allow it to be buried completely. 'Rereading the first chapter of the book of Genesis,' he wrote several years later, 'I was astonished to discover that it can be completely explained on my view, and much better, it seems to me, than on any other view.' In 1641, his pride overcame his caution and he published an amended version of the theory in a book called *Principles of Philosophy*.

The book was a pale shadow of the original. Desperate not

to offend the Church, he chose his words with such care, and wrapped each controversial idea in such lengthy disclaimers, that at times he almost contradicts what he has just said. Despite this, his message still comes through:

> Thus we may be able to think up certain very simple and easily known principles which can serve, as it were, as the seeds from which we can demonstrate that the stars, the earth and indeed everything in this visible world could have sprung.

This was the key to Descartes's theory. The entire universe was governed by a few simple laws. It was these 'laws of nature', and not the hand of God, that had converted the primordial matter of the initial chaos into the complex world around us.

The laws of nature, however, required time to do their work. According to Genesis, God had created the Sun, Earth, Moon and mankind in just six days; Descartes's universe took a little longer to build. Each step in his theory called for time. Time, first, for friction to break down the primordial matter of the initial chaos into three elements. Time for the lightest of these elements – a luminous dust – to be sucked into the centre of giant vortices where it formed the stars. Time too, for particles of the bulkiest element – a coarse, lumpy matter – to fall onto the surface of some of these stars, extinguish their light and form the planets. Although Descartes never said exactly how long his version of Creation would have taken, he conceded it couldn't have been achieved in less than 'the space of many years'.

While Descartes's theory pushed back the beginning of time, this passed largely unnoticed by scholars in his day, for his other ideas were far more controversial and would eventually play a greater part in extending the age of the universe.

Ultimately, his lasting achievement was not this small extension of time, but removing God from the day-to-day running of the world.

The common perception of God in the sixteenth century had been of a hands-on benevolent deity, taking day-to-day, minute-by-minute control over all aspects of the universe – sending plagues, locusts, floods and rainbows, performing miracles as and when necessary. Descartes swept all that away. For him, God's relation to the universe was like that between a clockmaker and his clock. God had set it in motion, but once started it continued to run without needing further intervention. The universe was governed, not by the hand of God, but by 'the laws of nature'. By introducing the idea that everything in the universe could be explained by a few simple rules, Descartes sowed the seeds of a new philosophy that would influence Newton, and prove a catalyst for the scientific revolution.

Throughout Europe, Descartes's works stirred debate. Not least at Cambridge University. Roger North, an undergraduate there in 1667, 'found such a stir about Descartes, some railing at him and forbidding the reading him as if he had impugned the very Gospel. And yet there was a general inclination, especially of the brisk part of the University, to use him.'

Among those of the 'brisk part' in Cambridge at this time were two men who would take Descartes's ideas in two very different directions: the young Isaac Newton (who read Descartes's work avidly) and Thomas Burnet, a fellow of Christ's College and an ordained priest.

Although largely forgotten today, for a few years at the end of the seventeenth century Thomas Burnet was far more famous than his illustrious Cambridge colleague. His fame, or perhaps more correctly, notoriety, stemmed from a book he published in 1681 called *The Sacred Theory of the Earth*.

Thomas Burnet.

The book, which offered the first great geological theory of the earth's formation, became a best-seller, stirred up a furious controversy, and raised questions that ultimately revealed that the earth was far older than anyone imagined.

Little is known of Burnet's character, but he was clearly interested in philosophy and greatly influenced by the work of Descartes. Unlike Descartes, however, he believed passionately in the truth of Scripture, and this particular combination of rationalism and religion became the driving force for his geological theory of the earth's history. Burnet believed that by adopting Descartes's rational approach he could show that the events described in Genesis had really happened: that Noah's flood had actually taken place, and that the Garden of Eden had once truly existed.

The evidence on which his theory would be based was nature. Nature, he argued, was the work of God – 'God's great book of the world' – and as such could not lie. 'We are not to suppose that any Truth concerning the Natural World can be an Enemy to Religion; for Truth cannot be an Enemy to Truth, God is not divided against himself.' He put his faith in the rocks and mountains of the earth. No need to search the monasteries of the Middle East for fragments of ancient text as Ussher had done; instead, 'those Pieces of most ancient History, which have been chiefly preserved in Scripture' could be 'confirmed anew, and by another Light, that of Nature and Philosophy'.

Burnet's theory was born from a moment of inspiration. In 1671 he travelled to Europe, acting as tutor to a young English aristocrat, the earl of Wiltshire. Such 'grand tours', as they became known, were considered a necessary part of a young gentleman's education and continued to be popular among European aristocrats right up to the nineteenth century. Typically a tutor would spend three to five years travelling through Europe with his pupil, visiting language schools, fencing masters, royal courts and natural curiosities. Along the way the master would be expected to instruct his pupil in the classics, history, law, etiquette and a host of other subjects deemed desirable for a man of noble birth.

The journey had a profound effect on Burnet. As he and his pupil crossed the Alps, he became overwhelmed by the chaotic vista of mountain peaks stretching before him. Mountains, at this time, were generally viewed as ugly, the visible sign of a world in decline, and it was not unusual for passengers in a carriage to draw down the blinds so they didn't have to look at them. But for Burnet, they aroused curiosity. 'The Sight of those wild, vast, and indigested Heaps of Stones and Earth,' he wrote, 'did so deeply stir my Fancy, that I was not easy till I could give my self some tolerable Account how that Confusion came in Nature.'

As Burnet stood and gazed at the Alps, he would naturally have assumed, like many of his contemporaries, that they had been created *after* the formation of the world. It was inconceivable that these jagged, twisted rocks were really part of God's original design. There was nothing harmonious or regular about them, no sign of the elegant symmetry you would expect from a Supreme Architect; they looked, to him, like a 'World lying in its Rubbish'. Yet, he wondered, if they weren't part of the original earth, how had they formed?

There appeared to be no force in nature that could have raised the earth to the height of the Alps. All the natural processes he saw around him did the opposite. Wind and rain eroded the summits of mountains; volcanoes and earthquakes shook their foundations: ' 'tis certain,' he wrote, 'that the Mountains and higher Parts of the Earth grow lesser and lesser from Age to Age; and that from many Causes, sometimes the Roots of them are weaken'd and eaten by subterraneous Fires, and sometimes they are torn and tumbled down by Earthquakes.'

If there were no forces in the contemporary world capable of creating mountains, then, the answer must lie in the past. Turning to the Bible, Burnet reread its description of Noah's Flood.

In the six hundredth year of Noah's life, in the second month, the seventeenth day of the month, the same day were all the fountains of the great deep broken up, and the windows of heaven were opened. And the rain was upon the earth forty days and forty nights.

One line jumped out at him: 'the same day were all the fountains of the great deep broken up'. It not only implied that water had poured out from beneath the earth, but suggested that the earth's surface had been broken in the process. Perhaps this upheaval had created the mountains.

While still travelling, he began to develop his theory. From rainfall measurements he estimated that a continuous down-pour lasting forty days and forty nights would produce only 160 feet of water, not enough to flood the earth to the depth mentioned in the Bible. Furthermore, when he calculated how *much* water was required to flood the whole globe, he discovered it needed eight times more than was available in the present-day oceans. This, he asserted, proved that the water of Noah's Flood must have come from an underground reservoir.

On his return to England, he continued writing; developing a theory to explain why this water came to be underground, and how it had created the mountains. Over the next nine years his ideas expanded into his famous work, *The Sacred Theory of the Earth*.

Burnet's book was a triumph of imagination. In vivid passionate prose he wove stories from the Bible together with evidence from nature to produce an elaborate but plausible theory, which, in its first stages, closely followed that of Descartes.

Like Descartes, Burnet created his earth – which had the structure of an egg – from a chaos of tiny particles. First, the heaviest particles floating in the chaos merged together to form a solid, spherical core – the yolk – onto which the lighter particles fell, surrounding the core with a layer of water – the white. The next step, the formation of the 'shell', occurred when oily particles separated out of the water and floated to the surface, where they combined with a fine dust falling from the air above to form 'a certain Slime, or fat, soft, and light Earth'. This, in time, baked hard to form the crust of a featureless planet.

Burnet's description bursts with colour. This is how he describes the early world:

[It] had the Beauty of Youth and blooming Nature, fresh and fruitful, and not a Wrinkle, Scar or Fracture in all its Body; no Rocks nor Mountains, no hollow Caves, nor gaping Channels, but even and uniform all over. And the Smoothness of the Earth made the Face of the Heavens so too; the Air was calm and serene; none of those tumultuary Motions and Conflicts of Vapours, which the Mountains and the Winds cause in ours: 'Twas suited to a golden Age, and to the first innocency of Nature.

The transformation of this smooth, featureless globe into a mountainous landscape was the most dramatic event in history. Over time, he explained, the sun heated the surface of the earth and the water below. As the earth dried out and cracked, steam from the heated water forced its way out through the crevices, enlarging them until the earth was no longer supported. Large chunks of the crust broke off and tumbled into the water below, flooding the globe. The gaping chasms left behind became the oceans, while huge pieces of rock, thrust upwards in the confusion, formed the mountains.

Burnet's delight in his theory shines through the whole work. He was confident that the only way the mountains could have formed was through a catastrophic flood similar to the one he described. The flood in the Bible, therefore, must have been a real event. His 'rational' explanation, he believed, quashed, once and for all, the persistent jibes of the atheists, and reaffirmed the truth of Scripture.

But Burnet had made a significant compromise. His description of the world's creation needed considerably longer than the six days mentioned in Genesis. This didn't worry him unduly. He assumed that the early verses of Genesis were allegorical, and needn't be taken literally. After all, he explained, hadn't St Peter stated in the New Testament that 'one day is with the Lord as a thousand years'? In which case, the six days of Creation might as well be regarded as 6,000 years. Before publishing, though, he sought a second opinion,

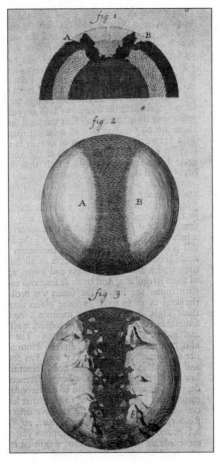

Burnet's explanation of the Flood. In fig. 1, the smooth crust of the earth cracks at A and B, and the two fragments 'fall in like Double Doors opening downwards'. As the crust collapses, water rises up from below to flood the planet (fig. 2). Finally, the water retreats leaving the earth in its modern form with continents and mountains on either side of a deep ocean (fig. 3).

and towards the end of 1680 sent a copy of the manuscript to Isaac Newton.

Newton, a devout Christian, immediately objected to Burnet's assumption that the account in Genesis need not be taken literally. 'As to Moses,' he replied, 'I do not think his description of ye creation either Philosophical or feigned.' Newton was adamant: if the Bible said it took six days to create the world, then that's exactly how long it took. This came as a shock to Burnet, who immediately sent another letter to Newton, seeking clarification. When Newton responded this time, it was with a typically ingenious solution: 'you may make ye first day as long as you please, & ye second day too,' he suggested, pointing out that at this stage in the Creation, the length of a day had yet to be determined. A day, he explained, was defined as the period of time it took the earth to make one revolution on its axis, and since, according to the Bible, the earth wasn't formed until the third day, before this time a 'day' could be as long as you want.

Newton also thought it likely that the newly created earth was stationary at first and only started spinning gradually. In which case, the first days of the world would have been much longer than they are today:

> And then if you will suppose ye earth put in motion by an eaven force applied to it, & that ye first revolution was done in one of our years, in the time of another year there would be three revolutions, of a third five, of a fourth seaven, etc. & of the 183d year 365 revolutions, that is as many as there are days in our year and in all this time Adam's life would be increased but about 90 of our years, which is no such great business.

In these suggestions, Newton showed that both time and the age of the earth could, in theory, be indeterminately long, and still agree with the biblical account. However, in practice these ideas had virtually no impact on the age-of-the-world debate; Burnet certainly didn't enlarge on them, preferring instead to stick to his initial premise that the account in Genesis was allegorical.

When Burnet published his *Sacred Theory of the Earth* in Latin in 1681, it was greeted with such enthusiasm that, at King Charles II's request, he immediately rewrote it in English. The English version proved even more popular. In the summer of 1684 the London diarist John Evelyn, returning a copy he had borrowed from Samuel Pepys, wrote: 'With your excellent book, I returne you likewise my most humble thanks for your inducement of me to reade it over again; finding in it severall things (as you told me) omitted in the Latine (which I had formerly read with great delight), still new, still surprizing and so rational, that I both admire and believe it at once.' Burnet rose in royal esteem, and on the accession of William III was appointed the king's principal chaplain and considered for the post of archbishop of Canterbury.

Despite this success, the book received mounting criticism from a small but vociferous group of clerics who disliked Burnet meddling with the Bible. 'May I not now conclude for certain,' wrote Herbert Croft, the bishop of Hereford, 'that this man hath been in the Moon, where his head hath been intoxicated with circulating the Earth, and is now come down to us with these rare Inventions.' He was even mocked in verse. In a ballad called *Battle Royal*, the satirist William King cast Burnet as an atheist and made him declare:

> That all the books of Moses
> Were nothing but supposes;
> That he deserv'd rebuke, Sir,
> Who wrote the Pentateuch, Sir,
> 'Twas nothing but a sham.

Change and Decay

The truth was that, despite these attacks, *The Sacred Theory of the Earth* was a great success; perhaps not with the most devout theologians or the most austere philosophers, but certainly with the public. People loved its flowing prose, colourful descriptions and compelling narrative. It was, in short, a ripping good read, and if perhaps eyebrows were raised at one or two of the details, its main theme – that the Flood of Genesis had shaped the modern world – became widely accepted.

But if Burnet thought he had established, without doubt, the truth of Scripture, he was wrong. By raising the question of how the earth was formed, his theory drew attention to a subject that had barely been considered before. His book could scarcely have met with a keener audience.

Science at this time was entering a golden age of inquiry. Natural philosophers were turning away from reading the dog-eared texts of ancient philosophers in favour of discovering the world anew through observation and experiment. The first scientific societies – the *Accademia del Cimento* in Florence, the *Royal Society* of London and the *Académie Royale des Sciences* in Paris – all sprang up in a ten-year period from 1657, and a new breed of wealthy gentry were emerging with the free time needed to carry out research. Before long, coffee houses and salons across Europe were ringing with the sound of passionate debate about the creation of the world.

Following the debate with interest were two British naturalists, John Ray and Edward Lhuyd. While both had read Burnet's theory and admired his ingenuity, neither could reconcile his account with their own observations in the field. As such, they were among the first people to glimpse the immensity of time.

Ray, tall and thin, was the son of an Essex blacksmith and the older of the two by some thirty years. Despite his poor background, he had received a scholarship to study at

Cambridge University, and later had ridden the length and breadth of Britain to complete a splendid and highly respected description of the flora of the country.

Lhuyd, in contrast, was a fiery Welshman who had been left penniless when his father backed the losing side in the English civil war. For years he had tolerated the menial work and measly pay of assistant keeper of England's first public museum, the Ashmolean in Oxford, until his appointment as curator in 1691 had given him the freedom to do what he wanted most – to travel. Ever since, like Ray, he had been exploring the botany and geology of Britain.

Surprisingly, given their long friendship, Ray and Lhuyd never met, but pursued their relationship with frequent letters. They were clearly absorbed in their work, as is evident from a letter Ray wrote to Lhuyd some ten years after they first began corresponding: 'you may possibly have heard, though I do not remember I ever told you, that I have four daughters'.

At the time, Ray and Lhuyd's pursuits were quite new and in some ways appeared rather strange to contemporaries. They were among the first naturalists to examine nature closely for its own sake. As they travelled from village to village, their peculiar activities – collecting stones, drawing flowers and asking questions about the landscape – drew suspicious stares from the locals, or in some cases outright hostility. In Wales, Lhuyd and his party were considered conjurors; in Cornwall they were arrested as thieves; and, on a trip to Brittany in northern France, Lhuyd was accused of being a spy and thrown into jail for two weeks. It obviously wasn't easy being a naturalist at this time. There is a story that Ray was summoned to give evidence in the unfortunate case of Lady Granville, a keen entomologist living in Exeter, whose relatives wanted her declared insane, and her will annulled, on the sole grounds that she collected insects.

Both Ray and Lhuyd firmly believed in the biblical account

of Creation, but they were also men who looked hard evidence in the face and didn't shrink from what they saw. From their observations, it was evident that if the earth had changed at all since it was created, it had changed only slowly.

One of the first pieces of evidence that pointed in this direction came from a trip Ray made to the Continent. In April 1663, while visiting Bruges, he noted in his journal that when workers in nearby Amsterdam had driven a well through one hundred feet of earth they had 'met with a Bed or Floor of Sand and Cockle-shells'. The presence of sea shells indicated that this must once have been the sea-floor, and Ray deduced that it had since been covered by sediment washed down by nearby rivers. But what puzzled him was how such a great depth of sediment could have accumulated in so little time. It was, he pondered, 'a strange thing, considering the novity of the World, the Age whereof according to the usual Account is not yet 5,600 years'. At the time, Ray was more interested in botany than the history of the earth, and thought little more about this observation. Thirty years later, however, his interest was reawakened by an intriguing letter from Lhuyd.

During a trip to study flowers on Mount Snowdon in his native Wales, Lhuyd had passed through Nant-Phrancon, a steep-sided valley to the north-east of Snowdon, where he noticed thousands of giant boulders lying scattered over the valley floor. He made inquiries and learnt from the locals that the last of these boulders had fallen just seven years earlier.

It happened in the valley of Nant-Phrancon, anno 1685, that part of a rock of one of the impendent cliffs, called yr Hysvaë, became so undermined ... that losing its hold, it fell down in several pieces ... and several stones were scattered at least 200 yards assunder. In this acci-dent one great stone, the biggest remaining piece of the broken rock ... continued its passage through a small meadow and a considerable brook, and lodged itself on the side [of] it. From hence I gather, that all the other vast stones that lie in our mountainous valley have, by such accidents as this, fallen down.

His curiosity aroused, he asked the nearby villagers how often such rocks fell, and discovered that only two or three had fallen within living memory. As he cast his eyes over the thousands of boulders littering the valley floor he was forced to conclude that, if only three had fallen in sixty years, then, 'in the ordinary course of nature we shall be compelled to allow the rest many thousands of years more than the age of the world.'

Lhuyd's observation forced Ray to confront a nagging doubt that had been growing in his mind since his trip to Bruges all those years before. Back then he had been astonished at the amount of sediment washed down by the rivers in such a short span of time. More recently, his concerns had focused on fossils.

Fossils were one of Ray's passions and a source of endless fascination. Nothing matched the thrill of removing them from the rock. After a day spent scouring the foreshore and cliffs of some promising beach, he would light a fire close to the water's edge and, gathering up that day's collection of stones, would place them into the heart of the flames. When the fire had died to glowing embers and the stones were thoroughly heated, he would lift them one by one with a pair of tongs and plunge them into a pool of cold water. Sometimes just a cloud of steam boiled off the surface and the stone was left intact; occasionally the fossil cracked in half and was ruined; with luck, however, the shock of the sudden quenching separated the fossil from its case and Ray would take the still warm stone from the pool and examine it in the light of the fire.

As he turned it over in his hands he would have wondered what it was, and how it came to be embedded so deep in the rocks. No one knew for certain the answers to these questions. For centuries philosophers had vaguely assumed that fossils had been left by the Flood but privately, Ray had his doubts. As the floodwater drained away, it would have deposited the fossils in a single layer, but the ones he found were buried at many different layers, deep in the earth. He also noticed that some groups of fossil shells were not widely scattered in chaotic heaps – as they would be if the water had been churned up by a flood – but lay together in neat rows, or beds. 'Such Beds,' he wrote, 'must in all likelihood have been the effect of those Animals breeding there for a considerable time, whereas the Floud continued upon the Earth but ten Months.'

He was further troubled by another mystery. While he recognised many fossils as being identical to shells he picked up on the beach, or fish caught in the ocean, there were some that appeared to be of unknown species. The strange coils called 'snake stones' (ammonites) that he found at Whitby on the north-east coast of England and the ferns preserved in coal which Lhuyd sent him from Wales had no living counterparts. Perhaps they were the remains of plants and animals now extinct.

At the time, this was a daring suggestion. Like most of his contemporaries, Ray believed that God had designed a perfect world; it was unthinkable that he would create a species only to destroy it again. Yet there seemed no other explanation. These fossils, he confided to Lhuyd, 'overthrow the opinion generally received, that since the first creation there have been no species of animals or vegetables lost, no new ones pro-duced'. If he was right and these were extinct species, then they must have needed time to live and die; 'there follows such a train of consequences,' he concluded, 'as seem to shock the Scripture-history of the novity of the world.'

Ray never pursued these ideas; indeed as a devout Christian

he found it difficult to make sense of the evidence that suggested the earth had changed only slowly. Many years later, fossils and strata would prove key pieces of evidence for a long-lived world, and although Ray glimpsed their significance, his faith in biblical chronology and his fear of speculation prevented him from publicising his discoveries.

In questioning the Flood, however, Ray was in a minority. At the end of the seventeenth century the overwhelming majority of naturalists believed that the geological evidence strongly supported the Bible's version of history. As long as that mind-set endured, the world would remain a few thousand years old.

5. God's Force

> In the year 1666 ... whilst he was musing in a garden
> it came into his thought that the power of gravity (w^{ch}
> brought an apple from the tree to the ground) was not
> limited to a certain distance from the earth but that this
> power must extend much farther than was usually
> thought. Why not as high as the moon said he to himself
> & if so that must influence her motion & perhaps retain
> her in her orbit, whereupon he fell a calculating what
> would be the effect of that supposition.
>
> – *from an account by John Conduitt,*
> *the husband of Newton's niece*

In the spring of 1687, at the age of forty-four, Isaac Newton
completed his masterpiece, *Philosophiae Naturalis Principia
Mathematica, The Mathematical Principles of Natural Philosophy,* commonly known from its Latin title as the *Principia.*
Nothing in science would ever be the same again.

It had taken the astronomer Edmond Halley, the book's
publisher, three years to coax the work out of the cautious
and highly suspicious Cambridge mathematician. But what a
work. Its breadth and scale went far beyond anything Halley
had dreamed of. The *Principia* swept away Descartes's largely
vague and undefined laws of nature and replaced them with
precise laws rigorously derived from mathematics, laws of

motion, of forces and of friction. But the brightest jewel in this glittering crown was Newton's concept of a force that acted at a distance – the universal law of gravitation.

We might assume that Newton's great work would immediately usher in a new era in philosophy, an era in which the old biblical account of the earth's history would be abandoned in favour of explanations based on his precise mathematical laws. Nothing could be further from the facts. Instead, his concept of gravity was immediately seized upon and turned to support the biblical account. Within ten years two philosophers, advancing two separate theories, used it to provide a 'rational' explanation of Noah's Flood. In doing so they extended belief in the biblical timescale well into the following century.

The first of these philosophers was John Woodward, the thirty-year-old professor of medicine at London's Gresham College. A careful observer of nature, Woodward believed that the Flood had played a crucial role in shaping the modern earth. While Burnet had originated this idea, it would be Woodward, with the help of gravity, who filled the gaping cracks in Burnet's hypothesis.

Woodward's rise to eminence appears to have stemmed from a stroke of good luck. While working as an apprentice to a London linen draper, his intellectual interests happened to be spotted by Peter Barwick, King Charles II's personal physician, who took him into his house and tutored him in medicine 'and all other useful learning'. He proved so able a student that within six years he was elected professor of medicine at Gresham College, at that time the nearest London had to a university. The position came with a clause in his contract forbidding him to marry, but this was of no consequence to the renownedly homosexual Woodward, and he lived happily at Gresham until the end of his life.

A man of bounteous energy and many interests, Woodward

John Woodward.

played a prominent and active part in the scientific life of the capital. Few philosophers, however, can have attracted so many disparaging comments from their contemporaries. To Lhuyd he was 'a proud, arrogant fellow', Ray thought him 'rude and insolent', and another acquaintance described him as 'vain, foolish and affected'. While it's possible that a homophobic attitude coloured some contemporary reports, it appears that many of these comments were justified. Woodward was thrown off the Council of the Royal Society for insulting Hans Sloane during a debate, and once, at the gates of Gresham College, he drew his sword and began to duel with a fellow doctor over the relative merits of vomiting or purging as a cure for smallpox. The duel ended embarrassingly when Woodward tripped and fell flat on his face – much to the delight of his detractors in the London press who reported the incident with glee.

Despite his quarrelsome reputation, many philosophers

respected Woodward as an innovative and methodical explorer of the earth. Inspired by the chance discovery of some fossils during a visit to the Cotswolds, he determined to learn all he could about these strange stones. So thorough was his research that by the end of his life his collection of fossils, meticulously labelled and carefully preserved in his rooms at Gresham, would be the finest in the land.

The most important of Woodward's insights was his recognition of the true nature of fossils. Even in the late seventeenth century, the origin of these lifelike objects was in doubt. Some naturalists thought they were stones that had formed naturally in the earth like crystals; others suggested they had grown from seeds washed by rain into the deepest crevices of the rocks. Woodward, however, recognised them as 'the real spoils of once-living animals'.

While this was not a new idea, his genius lay in providing an answer to what had previously appeared an intractable problem: if fossils truly were the remains of creatures drowned in the Flood, as many believed, then how did they come to be buried at so many different depths, so deep into the earth? A retreating flood would have left all the fossils in one layer near the earth's surface, not scattered through the strata.

Woodward's solution was to turn to gravity. Like Burnet before him, he proposed that the water for the Flood came from an interior ocean hidden beneath the earth's crust. In the normal state this was held in place by gravity, but, at the time of the Flood, God had momentarily suspended the full force of gravity and the water had spilled out. The effect was dramatic. Even with the partial gravity God left to prevent the earth completely disintegrating and flying off into space, 'the whole terrestrial Globe was taken all to pieces and dissolved'. The ordinary minerals and metals of the planet broke up into tiny particles which then dispersed in the vast ocean of water; while the remains of plants and animals, being organic, were held together by their fibres and stayed in one piece, floating about in the solution. When God restored the full force of gravity, everything descended according to its relative density: the heaviest rock particles and organic remains sank first, followed by the lighter material which sank more slowly. The result was that the heaviest fossils were buried in the deepest strata while the lightest settled nearer the surface. Thanks to gravity, Woodward had neatly explained why fossils were scattered through the strata.

His theory also cleared up another mystery: why European rocks contained the remains of creatures from warmer climates, such as elephants and tropical fish. Woodward explained that these remains had indeed originated in the tropics, but during the long Flood had drifted north on ocean currents until they reached Europe. At which point God had switched gravity back on, and they had sunk to earth.

When Woodward published his theory in 1695, the response of his British readers was at best lukewarm. Although his observations of fossils were regarded as authoritative, his mechanism for the Flood was greeted with incredulity. One wag dubbed it the 'hasty pudding' theory for the way the jumbled elements of the Flood were thrown together to form the earth. Nor, in the long term, was its reception helped by his personality. In later years Woodward became a favourite target for the London satirists. He was lampooned in pamphlets, and caricatured on stage as the comic Dr Fossile in the popular comedy *Three Hours after Marriage*. Submerged beneath the weight of this ridicule, his ideas struggled to gain an audience.

On the Continent, however, scholars were free from these prejudicial distractions, and here Woodward's theory received an enthusiastic welcome. Translated into Latin, French, German and Italian, it attracted the support of several European scholars who began to teach Woodward's ideas at their universities. The idea that fossils were the remains of living plants and animals that had died in the Flood began to spread.

Combined with the reports of explorers, this concept gave a tremendous fillip to the biblical account. Wherever they travelled, European adventurers found fossils. From the hills of China to the mountains of Peru, from Urswick to Uzbekistan. When they returned, they reported their discoveries to the newly formed philosophical societies, where the fossils were seen as evidence, not only that the Flood had really happened, but that Moses had been right – it really had covered the globe.

With ghoulish curiosity, however, the fossils that everyone most wanted to see were the remains of the humans who had drowned in the Flood. They were duly supplied. By the mid-eighteenth century no museum of curiosities was complete without its relic of the deluge. Italians wondered at 'The Stone Knee' of Verona, inquisitive Germans gaped at the 'Granite Man' of Worms, and even the collection of the Royal Society in London contained part of a 'human leg'. In Spain the naturalist Jose Torrubia proudly displayed some enormous bones which, he explained, were the remains of the 'giants' mentioned in the Bible; while in America Cotton Mather sent a letter to Woodward describing the bones and teeth of a similar antediluvian 'giant' discovered near Albany, New York. Not all such relics found their way into museums. In Valence, in southern France, the bones of a drowned 'giant' were hung in a church as a salutary reminder to the faithful that the Flood was caused by the sins of man.

Ironically, one of the first people to cast doubt on all this 'evidence' was one of Woodward's most ardent supporters, Johann Jacob Scheuchzer, professor of mathematics at Zurich University. A prolific author, whose wide range of interests included geography and meteorology, Scheuchzer also collected fossils with an enthusiasm and dedication to detail that made him one of the most respected naturalists in Europe.

God's Force

Possessing a detailed knowledge of human anatomy from his early training as a physician, Scheuchzer realised that the bones displayed as human relics were nothing of the sort. The famous 'Teeth of Giants' and 'Giants' skeletons' usually turned out to be the teeth and bones of an extinct elephant-like creature (later identified as the mammoth); while most other relics were merely stones that happened to bear a passing resemblance to parts of the human body. Given that virtually all the supposed relics belonged to other species, it became apparent that few, if any, remains existed of humans who had drowned in the Flood. Desperate to correct this stark anomaly, Scheuchzer began searching for genuine human remains.

In 1708 he made the first of two discoveries. During a walking tour of the hills fifteen miles east of Nuremberg, he stumbled across two fossil vertebrae beneath the gallows at Altdorf. Convinced they were human, he put them on display in his museum. For the next seventeen years, these two solitary fossils took pride of place in his collection as the only genuine evidence that humans had died in the Flood. Throughout this time he continued his quest for more evidence, but in vain. Until one day in December 1725 he made an incredible acquisition. On Christmas Day, in a rush of excitement, he dashed off a letter to Hans Sloane at the Royal Society in London describing his purchase: 'By a stroke of good fortune, there has been added to my modest museum a fragment set in fissile stone from Oeningen wholly deserving of thorough scrutiny, in which one can plainly discern, not the figments of capricious imagination, but numerous parts of the human head.' From a quarry in southern Germany he had acquired two relics: the aforementioned skull, and a complete head and torso of a petrified 'human' skeleton.

He was especially delighted with the second specimen, which he proudly put on display in his museum. On a label

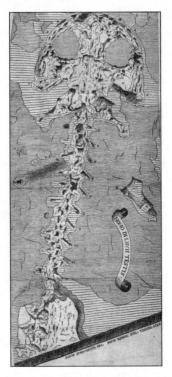

Scheuchzer's *Homo diluvii testis* – the Man who Witnessed the Flood.

beneath it he wrote: *Homo diluvii testis* – the Man who Witnessed the Flood. This was the final proof he needed, 'no one who would direct even a passing glance,' he declared, 'not to say an attentive study, upon this stone would fail to recognise it as a real and authentic relic of the Flood.'

Years later, *Homo diluvii testis* was acquired by the Teyler Museum in Haarlem, where in the early nineteenth century it was examined by the eminent French naturalist, Georges Cuvier. After chiselling away some of the surrounding stone to reveal the rest of the skeleton, he pronounced that it was nothing more than the remains of a large salamander. Likewise, the two vertebrae Scheuchzer had been so convinced were human turned out to be bones from a large fish.

Back in England, just a year after Woodward's book appeared, a young English clergyman named William Whiston published an even more ingenious explanation for the Flood. While this new theory also relied on gravity, it obviated the need for a miracle. Whiston, the twenty-nine-year-old rector of the fishing port of Lowestoft, on the east coast of England, had earlier studied mathematics under Newton at Cambridge, and while there had written an essay defending Burnet's theory of the earth. Inspired by Burnet's theory, and an observation of Edmond Halley, Whiston came to the conclusion that the Flood was caused by the gravitational pull of a comet.

Comets were a source of great excitement in the last decades of the seventeenth century; the inner solar system had received two dramatic visits from these 'bearded stars' in as many years. In 1680 an extremely long-tailed comet, known as the Great Comet, had circled the sun, followed two years later by the comet to which Halley gave his name. After observing both intently, and noticing how close to the sun they passed without breaking up, Newton declared that comets were 'solid, compact, fixed and durable, like the bodies of the planets'; while their tails were 'nothing but a very fine vapour'.

This was the starting point for Whiston's theory. He claimed that on 28 November 2349 BC a comet passed so close that the force of its gravity on the fluid beneath the earth's surface created an enormous subterranean tidal wave. So powerful was this gigantic wave, that when it crashed against the inside of the earth's crust it cracked it open and poured out to engulf the globe. But this was only the beginning. As the comet swept past, the attraction of the earth's gravity captured the long wake of water vapour lying in its tail. As the vapour condensed, it showered the earth with such torrential rain that, by the end of forty days, the planet was submerged beneath an ocean of water three miles deep.

Having dispensed with the need to summon up a miracle, the last part of Whiston's theory closely followed Woodward's. The earth, minerals and organic creatures suspended in the waters settled down according to their relative densities, the heaviest first and the lightest last, thus creating the strata visible today.

But the clinching detail that Whiston believed put his theory beyond doubt came when he identified the comet. The comet that caused the Flood, he claimed, was not just any comet; it was none other than that of 1680. Back in that year, the sight of this long-tailed comet had sent Edmond Halley scurrying to the history books for evidence of previous visits. From his search through ancient records he concluded that the same comet had been seen three times before: in AD 1106, AD 531, and September 44 BC – just a few months after the murder of Julius Caesar. Picking up on Halley's work, Whiston calculated that the comet would complete '7 revolutions in 4,028 years'. When he extended this cycle back, he found that the comet would have appeared in 2349 BC – the very year Ussher gave for the Flood.

Before publishing his theory, Whiston sought the advice of his old mathematics lecturer. He presented the manuscript 'before Sir Isaac Newton himself, on whose Principles it depended, and who well approved of it'. Newton not only approved of it, he was sufficiently impressed to summon Whiston from his parish in Lowestoft to be his assistant lecturer at Cambridge. Five years later, when he resigned as professor of mathematics, he named Whiston as his successor.

With three theories to choose from – Burnet's, Woodward's

and Whiston's – supporters of the biblical timescale were spoilt for choice. The fact that none of these ideas stood up to close scrutiny was generally overlooked. If the details weren't yet understood, it could only be a matter of time before they became clear, and then the Bible account would finally be confirmed. So, as the new century unfolded, naturalists moved away from devising elaborate theories and concentrated instead on making detailed observations in the field, working towards the day when nature would finally unveil her secrets.

While others drew on his theories, Newton himself never offered a scientific explanation for the Flood. But this didn't prevent him speculating on the timescale of the world. Surprisingly, for us, Newton probably spent more time and energy defending the chronology of the Bible than on any of his great scientific works. During the last thirty years of his life he made so many notes on theological and historical topics in his attempt to verify the biblical account, that by the time of his death the stacks of manuscripts ran to over a million words.

When his friend, Zachary Pearce, the rector of St Martin-in-the-Fields, visited him a few days before his death, he was still working on the problem:

I found him writing over his *Chronology of Ancient Kingdoms*, without the help of spectacles, at the greatest distance of the room from the windows, and with a parcel of books on the table, casting a shade upon the paper. Seeing this, on my entering the room, I said, 'Sir you seem to be writing in a place where you cannot so well see.' His answer was, 'A little light serves me.' He then told me that he was preparing his chronology for the press, and that he had written the greatest part of it over again for that purpose.

This, apparently, for the sixteenth time.

The purpose of all this effort was not to find the age of the world – for Newton broadly accepted the date determined by Ussher – but instead to re-date certain events in pagan history so as to make them square with the biblical account. What troubled Newton most was the date Ussher and others had given for the Greek capture of the city of Troy – the event that marked the rise of Greek civilisation. Ussher had placed this in 1184 BC, nearly two hundred years before the Israelites had built the first temple at Jerusalem. Newton thought this had to be wrong. It was unthinkable that the Greeks could have founded a great civilisation before God's chosen people, the Jews. In an attempt to correct what he believed was an obvious error, he began searching for a new date for Troy's fall.

Whatever his skills as a natural philosopher, Newton made a dreadful historian. Grappling with an age in which reliable sources were scarce and history had still to throw off the cloak of myth, he chose to base his chronology on the story of Jason and the Argonauts. He ignored the appearance in the narrative of six-handed giants, winged harpies, and armed men who sprouted from serpents' teeth, insisting that underneath this legendary tale was a real event – the journey of a Greek hero and his shipmates, who had rowed through the Dardanelles to become the first Greeks to enter the Black Sea. According to the records Jason had made his voyage just a few years before his compatriots captured Troy. By finding the date for this journey, therefore, Newton believed he could find the date for the capture of Troy.

While his source was suspect, his method was inspired. With no records of eclipses on which he could rely, he devised a new chronological technique based on the precession of the equinoxes. It had been known since the time of the Greek astronomer Hipparchos in the second century BC that the earth's tilted axis doesn't always point in the same direction, but moves slowly over a period of many years (about 26,000), like a wobbling top, to carve an imaginary circle in the sky. This *precession* means that the position of the stars appears slowly to change from year to year. Given an accurate description of the stars' position at any date in the past, it is possible to calculate how many years have elapsed since that time.

By piecing together information from several historical sources, Newton found what he believed was a description of the celestial sphere which the Argonauts were said to have carried for navigation. Marked on its surface were the positions of the constellations. From this, and knowing the rate of precession to be one degree every seventy-two years, he calculated that Jason and his crew had made their journey in 937 BC – nearly three hundred years after the traditional date. This was exceedingly convenient for Newton, for it postponed the siege of Troy and the rise of Greek civilisation to an era *after* the founding of the kingdom of Israel.

When Newton's *Chronology of Ancient Kingdoms Amended* was published a year after his death, few were convinced. While his astronomical method was sound, his treatment of historical sources had been nothing less than cavalier. He had ignored conflicting evidence, twisted fictions into truths, and bent the facts to fit a hypothesis that was intended solely to confirm his religious convictions. In addition, he had devoted his unique genius to a subject that was of rapidly fading interest.

A year after Newton was laid to rest in Westminster Abbey, he was joined by Woodward, who had arranged to be buried close to the only man whom he considered his intellectual equal. It was a curious conjunction of the two minds that had shaped the European view of the universe: while Newton's theories explained the heavens, Woodward's explained the earth. As Richard Bentley wrote at the time:

> Who Nature's Treasures wou'd explore,
> Her Mysteries and Arcana know
> Must high, as lofty *Newton*, soar,
> Must stoop, as searching *Woodward*, low.

While Woodward's ideas would soon follow their maker into oblivion, Newton's, as we shall see, would shortly inspire the first measurement of the earth's age based on the natural world alone. After Woodward and Whiston, few mainstream philosophers attempted to find confirmation of the Bible's narrative in nature. The world was studied for its own sake, and for its own truths, whatever they might be.

A glimpse of the possibilities of this approach came from Edmond Halley, who in 1715 became the first person to suggest that observations of nature could reveal the earth's age. He explained his thinking in a paper to the Royal Society. 'I shall take leave,' he announced, 'to propose an Expedient for determining the Age of the World, by a *Medium*, as I take it, wholly new, and which in my opinion seems to promise success.' All the oceans, he explained, contained salt, which originally came from the land. As rain percolated into streams, and streams flowed into rivers, the running water dissolved salt from the earth and carried it down to the sea. Halley knew from experiments heating pans of salt water, that when water in the oceans evaporated, the salt was always left behind. Over the centuries, the rivers of the world had continued to carry salt to the sea, making the oceans progressively saltier. '[W]e are thereby furnished', he declared, 'with an Argument for estimating the Duration of all Things.' By measuring the rate of increase of the sea's salinity, and assuming that salt had been washed into the oceans at a constant rate, it was possible to work backwards to a time when the sea contained no salt at all, which, Halley argued, would give the age of the world.

It was a brilliant concept, but, as Halley readily admitted, to carry out the measurement was beyond the technology of the day. The yearly rise in the sea's salinity was so small that even the best techniques were incapable of measuring it. He thought it would require readings taken centuries apart to register any change. '[I]t were to be wished,' he said, 'that the ancient *Greek* and *Latin* Authors had delivered down to us the degree of the Saltness of the Sea, as it was about 2000 Years ago.' If they had done so, Halley speculated, 'the World may be found much older than many have hitherto imagined.'

While his experiment would not be performed until two centuries later, Halley's idea struck an immediate chord. The possibility that there might exist a natural chronometer that could reveal the true age of the world excited the philosophers of the time. If ocean salt could not provide the answer, perhaps other methods could. In the middle of the eighteenth century, mankind's great quest to discover the scale of world history moved away from attempts to reveal the truth of Scripture, and instead concentrated on the search for a quantitative measure of the past. Nowhere was that search more intensive than in Paris.

6. The Heat Within

Georges-Louis Leclerc de Buffon, the director of the *Jardin du Roi* (now the *Jardin des Plantes*), watched keenly as the sun emerged from behind the clouds. Beside him in the grounds of the royal botanical garden in the centre of Paris stood the apparatus of his latest experiment: four large wooden frames holding between them an array of over a hundred and fifty mirrors. Screw-threads ensured that each mirror was precisely aligned to reflect the sun's light onto a plank of wood 158 feet away, the spot on which his eyes now focused. This was the test. Would the sun's rays be strong enough to burn the wood?

Buffon had been inspired to conduct the experiment by the famous story of how the Sicilian mathematician Archimedes had defended his home town from an attack by the Roman fleet. According to the legend, as the Roman galleys approached Syracuse, Archimedes had set up a giant mirror on the cliffs above the town, which he used to focus the sun's rays onto the ships' wooden hulls. One by one, the vessels had burst into flames, the fleet had sunk, and Syracuse had been saved.

The story had long been dismissed as fanciful – the great Descartes himself had said it couldn't be done – but Buffon was determined to prove him and everyone else wrong. In 1747, at the age of thirty-nine, he felt it was time to make his mark on the world, and a public demonstration of such a

Georges-Louis Leclerc de Buffon in 1761.

dramatic experiment would bring him the fame and admiration he craved.

Within two minutes the deal plank began to char, then to smoke. Just as it looked as if it would burst into flames, however, the sun disappeared behind a cloud. Buffon, though, had seen enough to declare the experiment a success. That summer, crowds flocked to the *Jardin du Roi* to see the now celebrated scientist set fire to buildings over two hundred feet away. Louis XV received a personal demonstration, and Frederick the Great of Prussia sent his congratulations. The flamboyant Monsieur de Buffon had made his mark on the world stage.

He would go on doing so for another forty years, sealing his reputation with two great works: the *Histoire naturelle*, a multi-volume epic in which he applied Newton's ideas to the history of the world to produce a new theory for the earth's creation; and *Les Époques de la nature*, in which he adopted the notion of a cooling earth to calculate the age of the planet. This final work dazzled the French public with a timescale so vast they could barely comprehend its enormity. But, as with all of Buffon's works, it was based entirely on solid scientific principles, and the evidence of nature.

According to contemporaries, Buffon was one of the most gifted, imaginative and hard-working men of his generation; he was also ambitious, vain and arrogant. Born in 1707, in Montbard, a small Burgundian town about a hundred miles south of Paris, he acquired the name Buffon at the age of ten when he inherited a small fortune from a great-uncle. His father, a local magistrate, shrewdly invested the inheritance in buying the nearby village of Buffon, thereby acquiring for his son both a name and – from the rent – an income for life.

From an early age, Buffon showed a flair for mathematics, and after Jesuit college and law school he travelled to Angers in western France to continue his studies. There he began to read the works of Isaac Newton, who proved a major influence. However, this early introduction to natural philosophy came to an unexpected end in 1730, when he found himself plunged into a quarrel over a woman with an officer of the Royal Croat regiment. In a town street lit by lanterns, the two men fought a duel. Buffon killed his opponent, and fled the town while the authorities dithered over issuing an arrest warrant.

Back in Burgundy he joined up with a young English aristocrat, the duke of Kingston, who was passing through the country with his tutor on a Grand Tour. The two became close friends, and travelled together through Switzerland and Italy, wining, dining, and visiting the theatre at every opportunity. While Buffon acquired a taste for the good life – one he never lost – he also learnt English. It proved a great asset. He could now communicate with English scholars, absorb the latest ideas from London's Royal Society, and read Newton's works in their original language.

In 1732 Buffon abandoned his hedonistic travels and settled down to a lifetime of industry. His mother's death the previous year had given him control of his own estate, and he used the money wisely. In a venture astonishing for the time, he acquired the hilltop fortress in the centre of his home town of Montbard, levelled the medieval buildings, and began constructing his own science park. It was a huge project, employing over two hundred people. Within the grounds of the old fortress, they planted a botanic garden, built a menagerie to keep lions and bears, converted an ancient tower into an astronomical observatory, and constructed the single-room cottage that would serve as Buffon's study for the next fifty years.

Once embarked on his scientific career, Buffon threw himself into his studies with single-minded determination. Certain

Buffon's hilltop estate at Montbard.

that success required hard work, he employed a servant named Joseph to drag him out of bed at five in the morning, with the promise of a crown every time he succeeded. One morning, unable to rouse his master by shouting, Joseph threw a basin of cold water over his slumbering figure. Buffon woke with a start, promptly got out of bed and duly rewarded him with his crown.

Even relationships took second place to his work. He later recalled: 'I had a little mistress whom I adored, but I forced myself to wait until six o'clock before going to see her, often at the risk of no longer being able to find her.' He could spare no time for long courtships: 'He looked only at young girls,' wrote a contemporary, 'not wanting to have women who would waste his time.' Another observer delicately remarked that 'he looked for his daily pleasure in a class of woman . . . who wouldn't take longer than the two minutes it is said the angels take to cover themselves with their wings, so as not to be jealous of our enjoyment.'

Within a short time, Buffon won himself a reputation as a savant. He carried out experiments on timber for the French navy, translated English scientific works into French, and wrote two famous papers on probability. This was the age of the Enlightenment, a time when people believed that reason could unlock the mysteries of the world, and Buffon's work was at its heart. In 1734 he was elected to the Academy of Sciences in Paris, and five years later, at the age of thirty-two, became director of the *Jardin du Roi*.

The *Jardin du Roi* was not only a botanic garden, it was one of France's leading scientific institutions. As well as greenhouses full of exotic plants, the grounds housed the king's natural history collection, and an impressive amphitheatre where the public could attend lectures and hear about the latest research. The position of director was highly coveted. 'All the medical world and all the Academy fought for that position,' wrote a friend. 'It is worth 1,000 crowns in salary, one of the most beautiful residences in Paris, and the right to make nominations for all the positions which depend upon it.' Buffon had found his home in the establishment; he would remain director for the rest of his life.

A visitor later remarked that Buffon 'always saw three things above all else: his glory, his fortune and his comforts'. His fortune was now guaranteed by his prestigious job; his comforts he found at his dinner table, and with his women; only glory eluded him. What better way to establish his name than to tackle the most fundamental philosophical question man could ask: When was the world created? And how?

Buffon was dissatisfied with the prevailing theories. After Burnet's *Sacred Theory of the Earth*, a string of books had emerged from England offering similar explanations for the world's landscape. Invariably based on the Bible, they nearly all had one thing in common: the conviction that the mountains were created by the Flood. Buffon dismissed them all. *The Sacred Theory of the Earth*, he remarked, was 'an elegant romance, a book which may be read for amusement, but cannot convey any instruction'. Unlike all previous theorists, Buffon chose to ignore the Bible. It would play no part in his theory, and he would make no concessions to its narrative.

Instead, his inspiration came from Newton. Buffon was a confirmed Newtonian – for years the only decoration on the wall of his study in Montbard was an engraving of the great man. Like many philosophers of his day, he believed that Newtonian mechanics would enable man to unravel the mysteries of the universe. If Newton's laws could explain the motion of the moon and planets, the rise and fall of tides, and even the shape of the earth, why shouldn't they also reveal our history? As Alexander Pope's intended epitaph proclaimed:

> Nature, and Nature's Laws lay hid in Night:
> GOD said, *Let Newton be!* and all was Light.

It was not surprising, therefore, that when Buffon began to develop his own theory of the world's creation, he should find his inspiration in Newton's masterpiece. In the *Principia* Newton had observed that it was a remarkable coincidence that all six planets (only Mercury, Venus, Earth, Mars, Jupiter and Saturn were known at this time) should rotate around the sun in the same direction. Intrigued, Buffon calculated that the odds against this happening, if each planet had been created independently, were 64 to 1. In addition, Newton had pointed out that the planets were all in the same plane of orbit – at least to within $7\frac{1}{2}$ degrees of each other – and the odds against this happening by chance were a staggering 7,962,624 to 1. 'It may therefore be concluded,' wrote Buffon, 'that the planets received their impulsive motion by one single stroke.'

The Heat Within

This was the starting point for his whole theory. While Newton believed the planets had been set in motion by 'the counsel and dominion of an intelligent and powerful being', Buffon believed the cause was a natural event – a collision with a comet. According to Buffon, at some time in the past a comet had smashed into the sun, sending a torrent of molten material streaming out into space. As this material circled the sun, its molten particles were pulled together by the attraction of gravity until they coalesced into spheres, forming the planets. The oblique angle of the initial impact imparted a rotational force which caused these newly formed, and still molten, planets to spin, some faster than others. Eventually, those spinning the fastest were unable to hold on to all their material and a small amount was thrown off to form moons. In this way, the single impact of a comet explained the creation of the whole solar system: how the planets were created, why they all revolved in the same direction, and why they all lay on the same plane.

Buffon's ideas were supported by observation. It had long been known that comets flew close to the sun, and often appeared to fall into it. Both the long-tailed comet of 1680 and Halley's famous comet two years later had passed extremely close to the sun – as Newton's calculations confirmed. The fact that they had emerged intact from the encounter implied that they were incredibly dense, possibly dense enough to dislodge some of the sun's mass should they hit it. And they didn't need to dislodge much. The mass of the planets was tiny in proportion to the sun's; just $\frac{1}{650}$ of its matter, Buffon calculated, would be sufficient to form them all.

These observations, though all very persuasive in their way, hardly added up to incontrovertible proof. There was, however, one piece of evidence that Buffon believed could put his theory beyond doubt – the shape of the earth. If the earth was made from the molten material of the sun, then it would not be spherical as most people assumed it to be, but oblate – flattened at the poles and bulging at the Equator.

Again, the idea came from Newton. Newton's laws showed that when a fluid mass of material came together, for instance to form a planet, the equal gravitation on all sides would, on its own, create a perfect sphere. However, if the planet was rotating when it formed, then centrifugal force would cause it to bulge at the Equator – the faster the spin, the bigger the bulge. When the planet cooled, it would solidify and retain the same shape.

If the earth was made from molten material dislodged by a comet, as Buffon proposed, then it had to be oblate. But was it?

The best evidence, collected half a century earlier, suggested that it was. In October 1672, Jean Richer, a French astronomer, had sailed to the South American island of Cayenne, just north of the Equator, to carry out some astronomical observations. While there, he noticed that the pendulum clock he had brought with him from France ran ever so slightly slower than it did in Paris. Although the difference was small, it was measurable. When he shortened the three-foot pendulum by just one twelfth of an inch, it corrected the error. Baffled as to why this should be, Richer re-checked the adjustment several times a week over the next ten months, and found it always remained the same. When he returned to France, he reported his findings to his colleagues at the *Académie Royale des Sciences*, who remained equally perplexed.

When news of Richer's discovery filtered across the English Channel to Isaac Newton, he immediately guessed the cause. The island of Cayenne, situated almost on the Equator, was further from the centre of the earth's gravitational pull than Paris. As the pendulum reached the top of its swing in Cayenne, the gravitational force pulling it back down would be slightly less than it would be in Paris; consequently, the pendulum would swing down at a slower speed. Richer's discovery revealed that the earth almost certainly bulged at the Equator; as such, it was also one of the first observational confirmations of Newton's theory of gravity. Meanwhile, for Buffon at least, it established that when the world formed, it was molten.

But none of this cut any ice with the old-guard philosophers of the *Académie*. Brought up on Descartes's theory of vortices, they distrusted the new-fangled Newtonian mechanics, with its assertion that space was a vacuum, and its strange force of gravity, whose power to act at a distance reeked of the occult. The French writer and philosopher Voltaire, who lived briefly in London, noted the different philosophical outlook on each side of the Channel:

> A Frenchman arriving in London finds things very different . . . He has left the world full, he finds it empty. In Paris they see the universe as composed of vortices of subtle matter, in London they see nothing of the kind . . . For your Cartesians everything is moved by impulsion . . . for Mr. Newton it is by gravitation. In Paris you see the earth shaped like a melon, in London it is flattened on two sides.

As long as French philosophers continued to believe that the earth was a perfect sphere, Buffon would find it impossible to convince them that his theory was correct. Fortunately, the young bloods of the *Académie*, led by Buffon's friend Pierre Maupertuis, were convinced that Newton was right, and they resolved to settle the issue once and for all. In 1735 they launched two expeditions. One, led by Maupertuis, headed for Lapland in the Arctic Circle; the other travelled to Peru at the Equator. In huge surveying operations, stretching over hundreds of kilometres, each team measured the length of one degree of the earth's meridian. When they returned to Paris some years later and compared their results, they confirmed that Newton was right: the earth did indeed bulge at the Equator. Buffon was ecstatic. This, surely, confirmed his theory.

Although Buffon was now certain he knew how the earth had formed, it would be another thirty years before he performed the experiments necessary to find its age. This delay is curious. When he wrote up his theory for the first volume of the *Histoire naturelle* in 1744, he must have known that he could use it to calculate the earth's age. Once again all he had to do was follow in Newton's footsteps. His speculations on the effect of the close encounter between the sun and the comet of 1680 showed the way:

> This Comet therefore must have conceiv'd an immense heat from the Sun, and retain[ed] that heat for an exceeding long time. For a globe of iron of an inch in diameter, expos'd red-hot to the open air, will scarcely lose all its heat in an hour's time; but a greater globe would retain its heat longer . . . And therefore a globe of red-hot iron, equal to our Earth, that is, about 40,000,000 feet in diameter, would scarcely cool in an equal number of days, or in above 50,000 years.

There was little difference between Newton's cooling globe of red-hot iron, and Buffon's cooling earth. Both would take a similar length of time to lose their heat, and Newton's initial calculation implied this would be at least 50,000 years. It would be relatively easy for Buffon to follow Newton's example and calculate how long the earth had taken to cool to its present temperature.

The reason he didn't, was because he assumed that if the earth *had* taken 50,000 years to cool to its present temperature, then this must have happened a long time in the past. Any cooling, he believed, had long since ceased, so there was no way of finding the earth's age. Much later, he discovered he was wrong.

Even without a new age for the earth, the completed *Histoire naturelle* included enough provocative statements to raise a storm. Most controversial was Buffon's starting point for his theory: his assertion that the earth was formed by a comet hitting the sun. In a stroke, he had reduced the creation of the world, the glorious masterpiece of the Supreme Architect, to nothing more than a catastrophic accident. The book's publication in 1749 created a furore. 'The devout are furious and want to have it burned by the executioner,' wrote an acquaintance. 'Truly, he contradicts Genesis in every way.' Hostile articles appeared in the religious press, and the theologians at the Sorbonne – who had the power to condemn works that contained 'principles and maxims not in accordance with those of Religion' – were forced to take action.

Buffon, however, was not some lowly author whose work could be condemned out of hand; as a respected civil servant and member of the *Académie* he had to be treated with circumspection. The theology faculty of the Sorbonne knew they had to handle the matter carefully. Too fierce a condemnation might force Buffon's resignation and inflame the scandal; too light, and they would appear to be endorsing his work. Discreetly, they sounded out his position before sending him an official letter:

Sir,
We have been informed . . . that you were prepared to satisfy the Faculty in regard to each of the articles it found reprehensible in your work; we cannot, Sir, praise you enough for such a Christian resolution, and in order to put you in a position to carry it out, we are sending

you the statements taken from your book that seemed
to us to be contrary to the beliefs of the Church.
We have the honour of being respectfully,
Sir,
Your very humble and obedient servants
The Deputies and Synod of the
Faculty of Theology of Paris
In the House of the Faculty,
January 15, 1751.

A list of fourteen 'reprehensible propositions' duly followed.
After delicate negotiations, it was agreed that the offending
passages could remain, but Buffon would publish a retraction
at the front of the next volume. 'I abandon everything in my
book respecting the formation of the earth, and generally all
which may be contrary to the narrative of Moses,' he wrote,
adding that he believed 'very firmly all that is told about Creation, both as to the order of time and the circumstances of
the facts.' None of this was sincere, but, as Buffon said later,
'It is better to be humble than hanged.' He had trodden a fine
line between compromise and condemnation and judged it
perfectly. His theory remained intact. So too did his career.

Although Buffon was reluctant to date the world without good
evidence, a fellow Frenchman had no such qualms. Some time
earlier, around the time he was preparing to write the first
volume of his *Histoire naturelle*, Buffon had received a copy of
a curious manuscript that had been circulating among Parisian
scholars since 1722. It claimed that the level of the world's
oceans had been falling for thousands of years. After measuring the rate of decline, the manuscript's anonymous author
had estimated that the earth was two billion (two thousand
million) years old. It was an astonishing claim – one Buffon
had dismissed as nonsense. However, he had been sufficiently

impressed by the author's arguments – if not his conclusion – to include them in his own work.

The manuscript is fascinating in its own right, because it reveals the difficulties philosophers had in drawing a coherent picture of the world from a mix of confusing evidence. As such, it is worth a short digression to explain its conclusions and how it came to be written.

The manuscript was titled *Telliamed*, and its anonymous author, Benoît de Maillet, was a French traveller and diplomat who had worked as the French Consul General in Egypt. It gave an account of a discussion that had supposedly taken place in Cairo between the author – described in the text as a French missionary – and Telliamed, an ancient and wise Indian philosopher. During the conversation, which lasted six days, Telliamed had revealed to the missionary how the earth was formed and why it must be at least two billion years old. In reality, this setting was fictitious: there was no missionary, and no Indian sage. Telliamed was simply the author's name spelt backwards.

The use of an oriental sage was a clever and not uncommon device for French writers of the time. It allowed them to put forward contentious, even heretical, views without being censored. If they valued their reputation, or feared condemnation, they would, like de Maillet, publish the work anonymously. However, by hiding his own name in plain sight, de Maillet could later claim credit should the book be a success.

In the first part of the story, Telliamed reveals how he knows that the sea-level is falling. Many years earlier, when he was just a boy, his grandfather had noticed that an island near the family's seaside home was always just submerged beneath the sea. Later in his life, however, he found that it always protruded above the waves. His suspicion that the sea-level was falling appeared to be confirmed when he explored some nearby hills and found sea shells embedded in the rocks.

Further confirmation came when his intrepid grandfather invented a diving bell and began exploring the bottom of the ocean. There he discovered the process by which mountains were made. Underwater currents, moving and depositing sediment, were building heaps and ridges of sand, silt and mud, creating a sea-floor which looked identical to the hills and valleys he found on land. If all the world's mountains were made this way, he reasoned, then the sea had once *completely* covered the earth.

But where had it gone? Telliamed explained that its fall was due to evaporation. Once water evaporated from the ocean surface, it rose through the atmosphere and dissipated into space. Eventually the earth would be left bone-dry. But this decline also offered the prospect of finding the earth's age.

To find out the rate at which the sea was shrinking, Telliamed's grandfather had built an elaborate measuring station on the coast near the family's home. Twice a year, in spring and autumn, a team of six philosophers opened an elaborate system of measuring wells linked to the ocean, and recorded the level of the sea. They had repeated this measurement every year for seventy-five years, slowly charting the sea's imperceptible decline. Their results revealed it was falling by three inches per century.

From the rate of decline, it was possible to deduce that the oldest seaports – which were now 6,000 feet above sea-level – had originally been inhabited 2,400,000 years earlier. Calculations further back in time were complicated, Telliamed explained, by the fact that when the oceans covered the whole world, their rate of diminution was less than now. Nevertheless, he deduced that the ocean-level had been falling for the past two billion years.

De Maillet's sources for this astonishing book were many and varied. The idea of the diminishing oceans originated with the ancient Greek historian Herodotus, and was supported by the eleventh-century Persian poet and astronomer, Omar Khayyam. In addition, de Maillet travelled throughout Egypt, collecting stories and observing the geology. One of the most startling pieces of evidence he gives for his theory is a description of the timbers of ancient ships he found lying in the desert near Cairo. The timbers are still there in the desert, only today the interpretation has changed. At Wadi-el-Faregh, to the west of Cairo, lie the fossilised trunks of prehistoric trees. Lying side by side, some of them look remarkably similar to the planking of a ship's hull.

Buffon borrowed several ideas from de Maillet. Like him, he believed that the mountains were formed by ocean currents, that the earth had once been completely submerged, and that the sea-level was still falling. But this last conclusion was also based on good authority. The evidence came from Sweden, where it was well known that ancient harbours, old anchors, and even wrecked ships could be found inland, far from the coast. In the 1730s and 1740s, Anders Celsius – the Swedish astronomer whose name is best known for the temperature scale – made a careful study of all the evidence he could find, and concluded that the water level was falling at the dramatic rate of four feet five inches per century.

For the benefit of future generations, he had a mark carved into a rock on the Swedish coast at Lövgrundet, near Gävle, showing the water level in his time. Today it is over two metres above sea-level. This is not, however, an indication that the sea-level is falling. During the last ice age Scandinavia was covered by an ice sheet several hundred metres thick. Like a boat with a heavy cargo, it sank into the earth's crust. Now that the ice has gone, the land is slowly rising.

None of this, of course, was known in Buffon's time, and Celsius's evidence persuaded him that the sea was falling. Furthermore, if it was falling in Sweden, it had to be falling everywhere. 'If similar observations were made in every country,' he wrote, 'I am persuaded that, in general, the sea would be found to have retired from every coast.'

But could he use this decline to calculate the age of the world? Buffon didn't think so. As a Newtonian, he knew that de Maillet's explanation – that the oceans were evaporating into space – couldn't be true. The pull of the earth's gravity prevented its atmosphere (which includes water vapour) from escaping. He also dismissed an alternative idea Celsius had put forward. Celsius had suggested that plants, while they were alive and growing, absorbed rainwater, which was then converted into soil when they died and decomposed. In this way, the earth's vegetation would steadily remove water from the oceans and turn it into land.

Buffon, however, had his own theory. He imagined that the original surface of the earth (the sea-bed), had been pock-marked with immense 'swellings and blisters', similar to the bubbles that appear in bubbling mud or molten metal – only millions of times larger. Over the course of time, the shock of earthquakes or the pressure of sea-water had progressively burst these enormous cavities. Each time a cavity burst, water rushed in to fill the empty space, and the sea-level dropped.

Unfortunately, the haphazard nature of this process meant that the sea's decline was unpredictable. In some ages it would drop faster than in others. As a chronometer of the earth's history it was useless, like a clock that ran fast on some days and slow on others. To Buffon, it must have appeared that the age of the world was destined to remain beyond reach.

For the next eighteen years, Buffon turned away from studying the history of the earth and concentrated instead on zoology. While he spent the winter months in Paris attending to the administration of the *Jardin du Roi*, at the first sign of spring he returned to his estate at Montbard, where he would spend the summer and autumn working on his books. His output was prodigious. Up every morning at five, he spent the first hour of the day responding to letters, then at six would walk up the hill, through his terraced gardens to the small study to begin writing. To the first volumes of his *Histoire naturelle* – on the theory of the earth and the formation of the planets – he added twelve more volumes on quadrupeds, then started work on another series, on birds. By the end of his life, his exhaustive study of nature would run to over thirty volumes.

Such was the life of a dedicated natural philosopher; though this picture of puritan hard work neglects the more colourful aspects of Buffon's personality. Vain in his personal appearance, he was reputed to employ a hairdresser two or three times a day, while on Sundays he paraded before the peasantry of Montbard dressed in the finest laced clothes.

As the summers rolled by and middle age crept up, Buffon finally settled down. At the age of forty-five he married Marie-Françoise de Saint-Belin-Malain, a poor, twenty-year-old student he had met during a visit to a convent in Montbard. She gave birth to a baby boy, Georges-Louis-Marie, soon nicknamed Buffonet, for whom his father had high hopes. In 1772 Louis XV elevated Buffon to the ranks of the aristocracy, and commissioned a statue of the new comte de Buffon to stand in the *Jardin du Roi*.

Despite his new-found status, Buffon still longed for fame and respect. In the early, heady days of his career, when he had impressed European scholars with his theory of the earth, and wowed royalty with his demonstrations of wood-burning mirrors, he had briefly tasted the adulation he desired. But since then, the public had lost interest. His diligently written volumes on the natural history of quadrupeds were not even selling enough copies to cover the cost of publication. His career was in decline. Instead of being an inspiration, the portrait of Newton mounted on the wall of his study had become a nagging reminder of all he had failed to achieve.

Then, in 1765, an announcement of a new discovery gave Buffon the opportunity he craved: the chance to re-enter the arena. It would lead to his greatest work, *Epochs of Nature*, and to a new date for the beginning of the world. The announcement came from Jean-Jacques Dortous de Mairan, a French mathematician, who revealed to the *Académie* that the earth contained an inner source of heat. The revelation galvanised Buffon. If the earth was still hot at its core, then perhaps it hadn't completely cooled after all. De Mairan's evidence was persuasive. He had compared the temperatures of the winter and summer months with the seasonal amount of sunlight, and found that winters were warmer than they would be if the sun was the only source of heat. As further evidence, he pointed out that the deeper one descended into a mine, the higher the temperature.

Buffon saw immediately that de Mairan's results fitted his theory. The heat preserved in the core of the earth was the residual heat from the time the earth had been part of the sun.

Suddenly, evidence he had earlier ignored fell into place. Housed in the natural history collection in the *Jardin du Roi* were numerous tusks and bones of 'elephants', hippos and rhinoceroses unearthed from the soil of northern Europe and Siberia. Many elaborate theories had been proposed to explain the presence of these tropical animals in the frozen wastes. Perhaps they had been driven north by the hunting expeditions of early man, or maybe they had died in the tropics, and their bones been washed to Siberia by a giant river running out of India. Buffon now offered a simpler explanation. The reason Siberia held the bones of tropical animals, he argued, was because it had once been as hot and as humid as Africa.

Armed with this new information, Buffon began writing *Epochs of Nature*, a history of the world from its conception to the arrival of man. It would be a radical work, not only

The Heat Within

because it pushed back the age of the world, but because it also proposed a rational explanation for the creation of animals, including man. According to Buffon, life had begun in the primitive oceans while the earth was still hot. Through chemical reactions, the heat had transformed 'oily and pliable' matter into organic molecules. Due to the intense heat, these first organic molecules had very quickly combined to produce complex organisms: the elephants, rhinos and giant ammonites whose remains were found in the northern regions.

Buffon even went so far as to hail Siberia as the cradle of life. In his new theory, life had formed on earth just as soon as the temperature of the once-molten planet had dropped to a tolerable level. This didn't happen everywhere at the same time. The equatorial bulge, he believed, acted as a store of heat, delaying the arrival of life there to some thousands of years after it appeared in Siberia. To Buffon, the frozen wastes of northern Russia were no longer the uncivilised edge of the world, they were the font of civilisation.

But far more exciting than this was the realisation that the answer to one of the fundamental mysteries of the universe lay within his grasp. If the earth was still cooling, then, by finding the rate at which it lost heat, he could calculate its age.

At first sight, the challenge appeared unsolvable. The rate of change in the earth's temperature was likely to be so small as to be immeasurable. Like Halley, attempting to measure the change of salinity in the oceans, Buffon may have wished that the Greeks and Romans had measured the temperature of the earth. Buffon, however, saw a way around the problem.

Throwing all his energy into the project, he set up a laboratory in a cellar on his estate at Montbard where he could model the effect of the earth's cooling. His idea was to heat tiny iron balls and then time how long they took to cool. He would then scale the results up to a sphere the size of the earth.

First, however, he needed to discover the relationship between the size of a sphere and how long it took to cool. This was essential. He couldn't simply assume that a sphere as large as the earth, and one the size of a cannonball, would both cool at a rate directly proportional to their diameters. Although Newton had assumed this was the case when he worked out that an iron sphere the size of the earth would take 50,000 years to cool, he confessed he didn't know for certain. 'I suspect,' he wrote, 'that the duration of heat may, on account of some latent causes, increase in a yet less ratio than that of the diameter; and I should be glad that the true ratio was investigated by experiments.'

To decide the issue, Buffon made a series of ten iron balls, with diameters ranging from half an inch up to five inches, which he heated in a furnace until they were white-hot. He then made two measurements: he timed how long they took to cool, first, to the point at which they could just be held in the hand, then when they reached the ambient temperature of the cool cellar. He dispensed with thermometers, as there was no practical way they could be used; instead he used his hands, believing that touch would give more accurate results. 'I let them cool without moving them, trying fairly often to touch them, and at the moment they no longer burned my fingers, and I could hold them in my hand for half a second, I marked the number of minutes that had passed since they were taken out of the fire; then I let them all cool to the current temperature.' To find the exact moment when the cooling spheres reached the temperature of the cellar, he compared them by touch with identical spheres that had been resting in the cool atmosphere for some time. To make sure both sets of spheres felt the same, the 'control' spheres had previously been heated in the fire to give them an identical blistered surface.

Buffon concluded that Newton's first guess had been right;

The Heat Within

the time the spheres took to cool *was* directly proportional to their diameter: the four-inch ball took twice as long as the two-inch. However, he disagreed with Newton's figure. From his results he calculated that a white-hot iron ball the size of the earth would take 96,670 years and 132 days to cool to the planet's present temperature – almost twice as long as Newton's estimate.

Because the earth is not made solely of iron, Buffon began a further series of experiments on a range of different substances representing the metals and minerals found in the earth's crust. Balls of gold, lead, silver, marble, clay, glass, chalk – over twenty different materials – were heated in the furnace and then cooled. A ball of solid tin placed in the kiln acted as a simple thermometer; when it started to melt, it was time to withdraw the balls.

This series of experiments, endlessly checked and repeated over a period of six years, became the subject of local gossip. According to one of Buffon's assistants, he employed four or five pretty women with very soft skin to accompany him into the cellar, where they would hold the cooling balls in their delicate hands, 'describing to him the degree of heat and cooling'. Buffon no doubt explained to the curious villagers that delicate hands were essential for the accuracy of the experiment.

Of the twenty or so different materials he tested, he used only the results of those most commonly found in the earth's crust: glass, sandstone, limestone and marble. Overall, he found these cooled quicker than iron. When Buffon completed his calculations he found that the earth would have taken 33,911 years to cool to the point at which it could just be touched, and a total of 74,047 years to reach its current temperature. He still wasn't happy with the results, however. He realised that all the time the earth was cooling, it was also receiving heat from the sun which would have lengthened the time it took to cool. With a determination for accuracy reminiscent of Ussher at his most pedantic he made one hundred and fifty pages of calculations to eventually conclude that the influence of the sun's heat merely added an extra 785 years to the age of the planet. In 1779 – three years after America declared independence, and ten years before France herself was thrown into revolution – Buffon published *Epochs of Nature*, in which he announced a new age for the world. The earth, he declared, was 74,832 years old.

Almost as dramatic as his date for the world was the shape of its history. It had taken nearly half the earth's age – 35,000 years – just for it to cool to the point when water could condense out of its atmosphere to form the oceans. Then, the first life-forms had appeared in the oceans. It wasn't until 60,000 years had elapsed, however, that the temperature dropped enough for the first land animals to appear – the elephants and rhinos that roamed the steaming jungles of Siberia. And it was almost 70,000 years before mankind appeared on the earth. 'Thus we are persuaded, independently of the authority of the sacred books, that man has been created last, and that he arrived to take the sceptre of the earth only when it was found worthy of his empire.'

While the publication of *Epochs of Nature* thrust Buffon back into the limelight, the response was not all he desired. Readers reacted with a mixture of admiration and incredulity. In short, they admired the style of the writing, but doubted the truth of the argument. It was, one reviewer wrote, 'One of the most sublime novels, one of the most beautiful poems that philosophy has ever dared to imagine.' Not a response likely to please the author of what was, after all, meant to be a work of science.

Ironically, the wave of disbelief that greeted Buffon's work was the result of living in a more rational age. In the thirty years since he had first originated the idea that the earth was formed by a comet hitting the sun, natural philosophy had moved on. Comets were no longer thought of as heavy, durable objects. The astronomer Joseph Lalande had recently stated: 'there is reason to believe that their substance has little density.' If this was true, a comet would be incapable of knocking a single planet out of the sun, let alone six. As if this wasn't enough to dampen confidence in Buffon's theory, there was also a better explanation for how the planets had formed. Twenty-four years earlier the German philosopher Immanuel Kant had published his *Universal Natural History*. While he agreed with Buffon that all the planets shared the same origin, he believed that this had been a vast cloud of matter rotating about the sun, not the collision of a comet.

Despite these setbacks, the work gained Buffon at least one passionate admirer – Catherine II of Russia. In gratitude for placing the frozen north of her country at the heart of civilisation, she sent Buffon a stream of furs, gold and jewellery. Buffon, whose vanity had not diminished with age, proudly returned the compliment by sending his son to St Petersburg to present the queen with a bust of himself.

In November 1779 *Epochs of Nature* was brought to the attention of the Sorbonne. It was a quarter of a century since Buffon had been forced to apologise for his theory of the earth. In the intervening years, the power of the theologians had withered while Buffon's position in the establishment had become impregnable. Besides, he was now too old to care what the divinity professors thought of his work. When he heard they intended to censure him once more, he wrote: 'I do not think that this affair will have any other regrettable consequence than that of causing a stir and perhaps requiring me to give the same foolish and absurd explanation they made me sign thirty years ago.' He was right. Once again he apologised, though this time he defiantly refused to publish a retraction.

Aeons

Although Buffon's estimate for the age of the world shocked the clerics, privately he believed the earth was considerably older than his published figure. Close observation of sedimentary rocks revealed that many of the most common types, including chalk and limestone, consisted almost entirely of tiny fossilised shells of sea molluscs whose life-forms could not have existed until after the formation of the oceans. According to his published account, the oceans had formed just 40,000 years earlier, when the earth had cooled enough for water to condense out of the atmosphere. However, when he looked at the thick layers of rock piled up, one above the other, and considered the origin of the organisms from which they were built, 40,000 years seemed a very short time. The molluscs would have taken many centuries to multiply to the enormous numbers required to produce such mountains of rock, while the process of decay – the grinding down of their shells, the deposit, drying, and hardening of their remains – would have taken 'not only centuries, but centuries of centuries'.

Buffon looked back at his calculation for the earth's age and began to adjust the figures. The initial experiment on the series of iron balls from which he had extrapolated his results had been carried out on such tiny spheres that in scaling up the results to the size of the earth there was plenty of room for interpretation. He believed he had been too conservative. In his unpublished manuscripts, he toyed with increasingly large numbers for the age of the world: first a million years, then three million, finally considering ten million years its most probable age.

Curiously, for someone so confident and opinionated, he never published these findings. This was not for fear of condemnation by the Church; ten million years was no more heretical than 75,000. The reason was partly due to a lack of evidence – these great ages were, after all, largely conjecture – and partly because he thought the public would find such a vast span of time incomprehensible. 'Why does the human mind seem to lose itself in the length of time . . . ?' he wondered. 'Is it not that being accustomed to our short existence we consider one hundred years a long time, and have difficulties forming an idea of one thousand, cannot even imagine ten thousand years, or even conceive of one hundred thousand years?' While he believed that 'the more we stretch time, the more we will near the truth', he knew that his readers were not yet ready for the truth. Regretfully, it was 'necessary to shorten it as much as possible to conform to the limits of our intelligence'.

After the publication of *Epochs of Nature*, Buffon continued studying and writing about the earth. Despite suffering from kidney stones which made the work increasingly painful, he never lost his fascination for the mysteries of the world. 'It is with regret,' he wrote shortly before he died, 'that I leave these interesting objects, these precious monuments of old Nature, that my own old age does not leave me the time to examine sufficiently to draw the consequences that I glimpse.'

In April 1788, Buffon died. The morning after his death, in accordance with his wishes, his body was cut open. Fifty-seven stones were found in his bladder, his heart was removed and presented to a colleague who preserved it in a gold and crystal urn, and his brain was measured and found to be 'of slightly larger size than that of ordinary brains'. Two days later, twenty thousand people lined the pavements and rooftops of Paris to watch as his coffin, pulled by fourteen horses and accompanied by sixty members of the clergy, passed through the city streets. That evening, six assistants from the *Jardin du Roi* carrying blazing torches accompanied the body for the first stage of the final journey to Montbard. There, a few days later, he was buried in a small crypt on his hilltop estate, a few yards from the study where he had spent most of his working life.

The following year, the French revolution erupted across the country; the crypt was broken into and the lead from his coffin stolen to make bullets. During the ensuing Terror, his son was arrested and condemned to death, in part by the very title his father had worked so hard to attain. Although, for most of his life, Buffonet had been a disappointment to his father, at the moment of his death he honoured his name with a touching tribute, full of courage and dignity. As he was led to the guillotine he turned to the crowd, and proudly declared: 'Citizens, my name is Buffon.'

In the last years of his life Buffon had stared deeper into the abyss than any man before. The vision he had glimpsed, of a past stretching back for millions of years, appeared as shocking as it was radical – a vision too startling to divulge to the public. Yet the public were not even ready to accept his shorter timescale; for most people, the biblical account of world history still ruled supreme. Over the coming years, however, as the inquiring minds of the Enlightenment sought a rational explanation for the structure of the earth, the popular perception of the age of the world began to change. Within sixty years, the once unassailable fortress of biblical chronology would lie in ruins.

7. Layers of Time

Patrick Brydone reined in his mule and looked out over the vast stretch of lava that lay before him. Black and barren, it was the last flow he had to cross before he reached the town of Acireale and a well deserved rest. In the distance, clouds of white smoke billowed from the snow-capped cone of Mount Etna. Behind the mountain, the afternoon sun was sinking fast. Giving the mule a sharp kick with his heels, Brydone pressed on, anxious to reach Acireale before nightfall.

Sicily was usually considered too dangerous to be part of the 'Tour'. The inns were said to be execrable – which was true – and the countryside was rumoured to be prowling with bandits – which was also true. Nonetheless, Brydone had heard such a wonderful account of the island from William Hamilton, the British envoy to the court of Naples, that he knew he had to see it for himself.

This was typical of his nature. Inherently inquisitive, Brydone had spent several years in Switzerland making experiments on electricity. For the last three years, however, he had earned his living by touring Europe, working as a bearleader, or travelling tutor, to a succession of students. It was in this role, on a warm May evening in 1770, that he found himself riding over the lava flows of Etna.

There were three in the party: Brydone himself; his young Scottish pupil, William Fullerton; and a friend, Mr Glover.

They had set out from Messina the previous day, and were now making their way down the east coast of Sicily, through some of the worst banditti country in the island. Fearing trouble, they had hired the services of 'two of the most intrepid and resolute fellows in the island . . . great drawcansir figures', as Brydone described them, 'armed cap-a-pie, with a broad hanger, two enormous pistols, and a long arquebuse' which 'they kept cock'd and ready for action in all suspicious places.' Thankfully, the guns and swords hadn't been needed; the worst obstacles they had faced were the old lava flows they had been picking their way through since Etna came into sight that morning.

This last one was the worst. 'I thought we never should have done with it,' wrote Brydone, 'it certainly is not less than six or seven miles broad, and appears in many places to be of enormous depth.' It had run down from the mountain many years before, and Brydone was impressed by the fact that, when it had reached the sea, 'it had driven back the waves for upwards of a mile.' He also noticed that, in contrast to the fertile countryside they had passed through earlier, scarcely a plant grew in the black clinker of the lava bed. It was so bare, he imagined that it must have been laid down in a recent eruption. He was wrong – as he would soon discover. The barren clinker they were riding over was ancient; what was more, it provided crucial evidence for the antiquity of the world.

The next day, the party arrived in Catania, the thriving harbour town lying at the foot of Etna. The city had recently been born anew. A century earlier a river of molten lava had poured down from the mountain, pierced its sixty-foot walls, and swept away half its buildings; thirty years later an earthquake had destroyed the rest. Now, in the fashionable style of the baroque, it was rising from the ashes.

Frustrated to find that there was no such thing as an inn,

Etna erupting in 1669.

Brydone called on a local priest for whom he carried letters of introduction, Canon Giuseppe Recupero. The canon turned out to be the perfect host. He found the party accommodation in a local convent, showed them around the city, and, best of all, enthralled them with his stories of local history. His extensive knowledge of vulcanology, and his facetious manner, immediately endeared him to Brydone, and before long the two were chatting enthusiastically about Etna and its eruptions.

Recupero turned out to be the local expert. He had become interested in the volcano quite by chance fifteen years earlier, when his abbot had asked him to investigate a stream of boiling mud that erupted from its side, and since then had made an exhaustive study of its geology. There was scarcely anything he didn't know.

He surprised Brydone by revealing that the lava field his party had crossed the previous afternoon came from no recent eruption, but had lain there for almost two thousand years. Its origin was well documented. According to the Sicilian historian Diodorus Siculus, it had flowed from the mountain in the third century BC, during the second Punic war. At that time the war was at its height, and the southern city of Syracuse was under siege by the Roman army. To relieve the siege, the Carthaginian commander, Imilco, began a forced march down Sicily's eastern coast. At Acireale – where Brydone had encountered the barren lava – Imilco found the route blocked by 'rivers and streams of fire', and was forced to make a lengthy detour around Etna. Recupero had confirmed this historical account from inscriptions he found carved into Roman monuments on the lava itself.

But the discovery that the lava field was close to two thousand years old had an even more dramatic implication. Taking Brydone into his confidence, Recupero revealed that Etna had to be older than the Bible said it was. As evidence, he led Brydone to a deep well which had been sunk through several layers of lava. What was so important about the well, Recupero pointed out, was that between each layer of lava lay a substantial layer of soil. From the barren clinker bed at Acireale, he knew that it took more than two thousand years to form even a thin layer of soil on the surface of a lava flow, therefore there must have been more than that space of time between each of the eruptions which formed the strata in the well.

Not far away was a deep pit that had been cut through *seven* such layers. 'Now', said Recupero, 'the eruption which formed the lowest of these lavas, if we may be allowed to reason from analogy, must have flowed from the mountain at least 14,000 years ago.'

It was an astonishing announcement, coming as it did from a Roman Catholic priest. In his letters, Brydone recalled that Recupero was exceedingly embarrassed by his discovery. 'Moses hangs like a dead weight upon him, and blunts all his

Patrick Brydone in later life, resting his feet after his travels.
Note the map of Sicily on the wall.

zeal for enquiry; for that really he has not the conscience to
make his mountain so young as that prophet makes the world
... The bishop, who is strenuously orthodox – for it is an
excellent see – has already warned him to be upon his guard
and not to pretend to be a better natural historian than
Moses.'

Back in England Brydone collected together the letters he
had written during his tour and in 1773 published them in a
book, *A Tour through Sicily and Malta*. It was a huge success,
immediately becoming one of the most popular books of its
day – largely because Brydone wrote candidly about everything he saw and everyone he met. Inevitably he included
his conversation with Recupero. 'What do you make of
these sentiments from a Roman Catholic divine?' he asked,

rhetorically. The reply from his readers was unequivocally hostile. 'Shall all the accumulated evidence of the history of the world; – shall the authority of what is unquestionably the most ancient writing, be overturned by an uncertain remark such as this?' fumed the great man of English letters, Doctor Johnson, while the bishop of Llandaff dashed off a fiery riposte saying it was insufferable for 'a minute philosopher to rob us of our religion'.

In later editions of the book, no doubt under pressure from his peers, Brydone's publisher felt obliged to add an apologetic footnote beneath the offending paragraph:

> This passage has been the subject of much severe comment. That Mr. Brydone, in putting into a whimsical light many of the religious superstitions of the countries through which he travelled, was incautious respecting the essential things of religion, is evident: it was an error into which a young man of sprightly talents and thorough devotion to natural science was then very apt to fall.

Word of Brydone's book eventually got back to Sicily where, according to the British consul general, it upset the local population. Some unnamed priests with whom Brydone had spoken were reported to have been denied promotion, while British travellers presenting letters of introduction were now turned away because Sicilians were too afraid to speak to them.

In an age when the European public still believed in the literal timescale of the Bible, Brydone's *Tour*, translated into French and German, and running to at least eleven editions in English, spread the news that the world might not be so young as people thought. Remaining in print for forty years, and in strong demand at public libraries, it reached a far wider audience than the works of Buffon or de Maillet. As such, it sowed further seeds of doubt in the accuracy of Ussher's chronology. As the poet William Cowper recorded in *The Task*, written in 1785.

> ... Some drill and bore
> The solid earth, and from the strata there
> Extract a register, by which we learn
> That he who made it, and reveal'd its date
> To Moses, was mistaken in its age.

However, it would take more than a single observation for the public to relinquish the familiar and comforting chronology of the Bible. It would take a revolution. Only by understanding the true process by which fossils came to be buried in the earth and raised to the heights of mountains would it be possible to grasp the enormity of geological time.

The question of the true origin of fossils was very much in the air in the 1760s. By this time, most rational thinkers had dismissed Woodward's and Whiston's theories as pure fantasy, but were struggling to find something better to put in their place. Typical of those seeking an answer was Josiah Wedgwood, the founder of the Staffordshire pottery business:

I should be glad to know from some of you Gentlemen, learned in Natural History & Philosophy, the most probable theory to account for these vegetables (as they once were) forming part of a stratum, which dips into the Earth to our knowledge 60 or 100 yards deep, & for ought we know to the Centre! These various strata, the Coals included, seem from various circumstances to have been in a Liquid state, & to have travel'd along what was then the surface of the Earth; something like the Lava from Mount Vesuvius ...

But I have done. I have got beyond my depth ... These wonderful works of Nature are too vast for my microscopic comprehension. I must bid adieu to them for the present, & attend to what better suits my Capacity. The forming of a Jug or Teapot.

Wedgwood was asking the right questions, but his suggestion that the strata of the earth had originated as lava-like liquids running downhill could not have been further from the truth. The solution to this mystery would not emerge from Staffordshire, but further north, in the cold and smoky city of Edinburgh.

'Your Scottish sun,' joked one Mediterranean visitor, as he peered through the smog at the pallid disc in the sky, 'is like our Sicilian moon.' The atmosphere may have been grey, but the mid-eighteenth century marked a golden era for Scottish intellectuals. The Enlightenment was at its peak. The classes in science and medicine at Edinburgh University were considered the best in the world, and its liberal attitude provided a welcome alternative to the religious restraint of the English universities. At this time, Oxford and Cambridge still awarded degrees only to confirmed members of the Church of England, so Nonconformist Englishmen as well as Calvinist Scots flocked to the 'Athens of the North', attracted by an enlightened philosophy that put reason before tradition. By the end of the century Edinburgh had more students than Oxford and Cambridge combined.

The intellectual excitement of the university rubbed off on the town. At its heart was the Oyster Club. Once a week a group of the city's brightest men gathered together to discuss the latest ideas, down tankards of porter, and feast on the fashionable local delicacy of oysters. All the bright lights were there: the economist Adam Smith, the chemists Joseph Black and Sir James Hall, and an unassuming gentleman farmer named James Hutton.

'It was always true of Dr Hutton, that to an ordinary man

he appeared to be an ordinary man,' remarked one of his friends after his death. And indeed, nothing in Hutton's dress or manner indicated the grandeur of the scheme that would later earn him the title, 'the founder of modern geology'. He wore simple clothes like a Quaker, possessed no airs or graces, and enjoyed the company of his friends like the next man. Yet this 'ordinary man' developed a very extraordinary idea.

It began almost by accident. On his father's death Hutton had inherited a small farm near Duns, forty miles south-east of Edinburgh, which he resolved to turn into a model of modern practice. Travelling to Norfolk, where farming techniques were said to be the best in the country, he spent a year living on a farm learning the latest methods of husbandry. During this time, his curiosity piqued by his interest in agriculture, he began asking questions about the landscape around him. How had it formed? Why did different regions have different soils? Where did the rocks come from? Soon he was 'studying the surface of the earth . . . looking with anxious curiosity into every pit, or ditch, or bed of a river that fell in his way'.

In 1768 with the improvements to his farm complete, he leased it to a tenant manager, and moved to Edinburgh. The move freed more time for his investigations, and over the coming years he travelled the length and breadth of Britain, building up a picture of its landscape. Few of Hutton's letters survive, but those that do paint a colourful picture of his travels. In 1774 he arrived in Bath, after a painful journey through southern England.

Lord pity the arse that's clagged to a head that will hunt stones . . . but how a pocket or a breek [pair of breeches] should not wear when mine in travelling less than 40 days has been more than 4 times mended – my arse, it is evident, is now a part of much greater consequence than my head, for a barber never looks at me whilst here a taylor is my constant companion – I carried one with me in a chaise to Wiltshire and a Taylor is the only acquaintance I have at Bath – no joke, dead earnest – It is a great advantage to a naturalist to have no acquaintance in a place provided he is to stay, for he must walk about & enquire to prevent hanging himself thro the day and then at night he writes a bit and drinks a soup [sup] of quiet toddy . . .

. . . I have this day packed a hogshead of bibles [geological specimens] all wrote by Gods own finger, they are to go by way of Greenock if they can find the way . . . I think I know pretty well now what England is made of.

By this time he had begun to formulate a theory. Wherever he travelled, he could see that rocks were decaying. The frosts of winter, the trampling of feet, the pounding of the rain; all slowly wore away at their surface. To Hutton this decay was wondrous, God-given even; it produced the soil on which man could grow his crops. But from the hours spent watching dirty floodwater flowing down streams and rivers, he could also see that the whole surface of the earth was ever so slowly being ground away and flushed out into the ocean.

This much was common knowledge. For years Christians had used the erosion of rocks as an argument against the atheistic idea that the earth had existed eternally. If the earth had existed for ever, they argued, by now the mountains would have eroded away to nothing.

Hutton, however, realised that this wasn't necessarily so. The erosion of land was only part of the story: just one stage in a lengthy cycle. After the tiny particles of rocks and minerals reached the sea, they settled as sediment on the ocean floor, slowly building up into layers hundreds of feet thick. Eventually the layers solidified into rock, which was then pushed up above the ocean to form land. The fossil shells found high on mountains, and in the chalk and limestone strata, were not left there by a retreating ocean, as de Maillet and Buffon had believed, but had been lifted up from the sea-floor by tremendous forces acting over a very long time. The mountains were maintained by an endless cycle of erosion and uplift.

But what could raise the sea-floor to the height of mountains? Hutton believed the answer was heat. By the mid-eighteenth century, the expansive power of heat was a well-known phenomenon. Naturalists had found that when they heated various substances – whether they be solid, liquid, or gas – nearly all expanded. And heat was clearly powerful: the new engines Hutton's friend James Watt was building in Birmingham depended on the pressure of steam created by it. In addition to this circumstantial evidence, there was also the more direct evidence of volcanic activity. The only places where land was known to be rising all lay in the vicinity of volcanoes, whose heat, Hutton assumed, was the driving force behind this uplift. Heat was a fair guess, but Hutton had yet to find the evidence to clinch his argument.

In 1785, while he was still accumulating evidence for his theory, Hutton was persuaded to give a paper to the newly formed Royal Society of Edinburgh. He chose the opportunity to reveal his ideas to the world. On 4 April, he stood up in front of the fellows of the society's Physical Section, and began to explain his theory of the earth.

Right from the start, his emphasis was on time. The aim of his talk, he explained, was 'to form some estimate with regard to the time the globe of this Earth had existed, as a world maintaining plants and animals'. He then moved on to explain his theory. For Hutton, the earth behaved like a dynamic machine. The gradual erosion of the land, its deposit as sediment on the ocean floor, its consolidation into rock and its subsequent uplift to form new land were all stages in a continuous cycle which had only one end in sight: to make the earth fit for life, specifically human life. While he broke with the literal interpretation of Genesis, he did not break with religion. Hutton's earth-machine was self-perpetuating, life-giving, the work of a benevolent God. In keeping with the enlightened thinking of the time, there was no place for the old idea of a wrathful deity who intervened in nature and punished the world with a flood. 'General deluges form no part of the theory of the earth,' he announced, 'for, the purpose of this earth is evidently to maintain vegetable and animal life, and not to destroy them.' Everything worked towards this end. The processes of erosion and decay were not malevolent acts of destruction, but beneficial parts of God's design, for by creating soil for plants to grow, they provided food for animals to live. Even volcanoes were beneficial, for they acted as pressure valves, releasing heat into the atmosphere 'to prevent the unnecessary elevation of land, and fatal effects of earthquakes'.

Having outlined his system to the audience, Hutton returned to his main theme: the length of time it had taken to build the world. 'What has been the space of time necessary for accomplishing this great work?' he asked. It was clearly long. 'If we could measure the progress of the present land, towards its dissolution by attrition, and its submersion in the ocean, we might discover the actual duration of a former earth. But how *shall* we measure the decrease of our land? Every revolution of the globe wears away some part of some rock upon some coast; but the quantity of that decrease, in that measured time, is not a measurable thing.'

Even if it could be measured, there was still no way to measure the earth's age. Because heat continued to raise the land as it was eroded, cycles of erosion and uplift could have continued indefinitely. How many times the earth had been worn away, only to be raised again to the surface, was impossible to say. 'The result, therefore, of our present enquiry,'

Hutton famously concluded, 'is that we find no vestige of a beginning, – no prospect of an end.'

At the time Hutton's theory was not widely accepted. To many, his concept that the earth had passed through endless cycles of erosion appeared dangerously similar to the pagan belief that history consisted of endless cycles of time, endlessly repeated in an eternal world. The clergy, and even other geologists, railed against him. When he valiantly responded with a two-volume work in which he expanded on his theory, those who read it were left so confused by his awkward style that it did little to help his case. As one reader commented: his explanations were set out in a 'manner so peculiar, that it is scarcely more difficult to procure the secrets of science from Nature herself, than to dig them from the writings of this philosopher.'

By now Hutton was in his sixties. If posterity had relied on his writing alone, his ideas might have died with him, but, fortunately for science, his enthusiasm for his subject rubbed off on his friends. His lively and engaging conversations infected them in a way his tangled prose never could, and in the last years of his life he gathered around him a group of disciples who would pass his ideas on to future generations. At the time of his death in March 1797, the diluvial theory – that the earth had experienced a tremendous flood – still held sway. But by then, Hutton had gathered enough followers to spread his message.

At the end of the eighteenth century, the dichotomy between science and the Church was as great as ever: on the one hand, the brief span of Ussher's chronology, on the other the indefinite epochs of Hutton's cyclic world. The two were incompatible, but in their search for truth the more emboldened rationalists of the Enlightenment had no regard for the consequences. 'Should free inquiry lead to the destruction of Christianity itself,' wrote the English chemist Joseph Priestley, 'it ought not, on that account, to be discontinued; for we can only wish for the prevalence of Christianity on the supposition of its being true; and if it fall before the influence of free inquiry, it can only do so in consequence of its not being true.' As more evidence emerged from the rocks, the room for different interpretations of the earth's past narrowed. The inevitable clash between science and religion was drawing near.

8. A Fossil Clock

The new century opened with a burst of activity. Between 1800 and 1840 the sound of hammer on stone rang out across Europe as geologists began exploring the earth in unprecedented numbers. More than ever before, this was a time when natural philosophers took to naming and ordering the objects they discovered, and not just the objects. The words *geology*, *biology* and *scientist* were either coined or came into popular use during this period. The growing interest in the natural world, particularly geology, also reflected the spirit of the age. This was the Romantic era – a time when poets and artists reacted against the growing industrialisation of society, and turned to nature for inspiration. The works of Wordsworth, Coleridge and Turner evoked the beauty and drama of the landscape, with its soaring mountains, thundering waterfalls and tranquil glades. The long tradition of Enlightenment rationalism and this new Romantic fascination with nature came together in geology, and in Britain especially, the subject became hugely popular. By 1807, the country had its own Geological Society, and hundreds of eager and often knowledgeable amateurs contributed as much to the science's progress at this time as the few salaried academics. What they uncovered in the first half of the century left no one in any doubt about the vast scale of geological time.

The most significant development during this period – and

vital to the understanding of geological time – was the identi-
fication and naming of the successive layers of strata. For
centuries, the chaotic nature of the rocks – in Byron's words,
'thrown topsy-turvy, twisted, crisped, and curled' – had prevented naturalists grasping that there existed an underlying
order. Knowledge of the different types was rudimentary. A
typical classification in use at the beginning of the century
divided them into just three categories. Hard crystalline rocks,
like granite, that generally seemed to underlie all the others,
were known as Primary rocks. Above them, the layers of softer
rock, such as sandstone, limestone and chalk (the ones we
now call sedimentary), were known as Secondary rocks. While
in the uppermost layer, above the chalk beds that formed the
last of the Secondary rocks, came another sequence of strata,
known as Tertiary rocks.

It was only when the English canal surveyor William Smith,
and the French naturalist Georges Cuvier, discovered that
each layer of rock had its own distinctive fossil fauna, that the
true order of the strata became apparent. The fossils changed
– for the most part, gradually – from one layer to the next,
and could be used, like the page numbers of a book, to identify
each layer of strata and determine its position relative to the
other layers. It didn't matter how folded or crumpled the
strata were; once the order of fossils was known, their correct
sequence could be unravelled.

As early as 1812, Cuvier, working at the Muséum National
d'Histoire Naturelle in Paris, constructed a rough sequence
of the animals that had lived and died in the past. First, and
most intriguing, the lowest layers of rock contained no fossils,
and appeared to portray a time before life existed. Then,
working up through the strata, the first simple forms of life
appeared: sea creatures – small zoophytes and molluscs – followed by plants, fish and freshwater reptiles. Later, large
turtles and lizards emerged, then the extraordinary creatures

A Fossil Clock

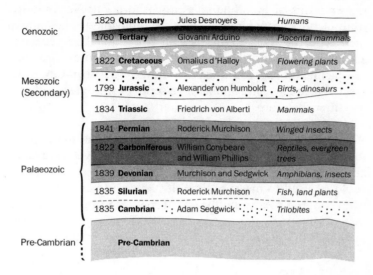

Cenozoic	1829 **Quarternary**	Jules Desnoyers	*Humans*
	1760 **Tertiary**	Giovanni Arduino	*Placental mammals*
Mesozoic (Secondary)	1822 **Cretaceous**	Omalius d'Halloy	*Flowering plants*
	1799 **Jurassic**	Alexander von Humboldt	*Birds, dinosaurs*
	1834 **Triassic**	Friedrich von Alberti	*Mammals*
Palaeozoic	1841 **Permian**	Roderick Murchison	*Winged insects*
	1822 **Carboniferous**	William Conybeare and William Phillips	*Reptiles, evergreen trees*
	1839 **Devonian**	Murchison and Sedgwick	*Amphibians, insects*
	1835 **Silurian**	Roderick Murchison	*Fish, land plants*
	1835 **Cambrian**	Adam Sedgwick	*Trilobites*
Pre-Cambrian	**Pre-Cambrian**		

The Golden Age of Geology. By 1841 geologists had named nearly all the major classifications.

that the English naturalist Richard Owen would later call dinosaurs. After these came more reptiles, vertebrate fish, rodents, birds and a profusion of mammals the like of which had never been seen before – or since. Close to the surface lay the remains of more familiar mammals: mammoths, rhinoceroses, horses. There were no fossils of human beings.

As well as being a guide to the order of the strata, these creatures posed their own mystery. Where had they come from? In 1802, the French naturalist Jean-Baptiste Lamarck suggested that the earliest organisms had formed by spontaneous generation, through the action of heat, sunlight and electricity. All subsequent life-forms, including man, he argued, had developed naturally from these first creatures. There was no extinction, just a gradual progression towards higher forms of life. The idea provoked a vitriolic response from clergy and naturalists alike. 'Abominable trash,' fumed the keeper of shells at the British Museum. The generally accepted view, and the one held by most naturalists at the universities of Oxford and Cambridge, was that God had created the plants and animals in a series of separate creations, one after the other.

Whatever their origin, there was no denying that, in the past, the earth's fauna had changed considerably. As ever, the realisation that change had taken place raised the possibility that it could be used to measure time. In the same way that Halley had proposed measuring the age of the earth from the change in the salinity of the oceans, and Buffon from the change in the earth's temperature, a softly-spoken, hesitant young barrister named Charles Lyell now attempted to use fossils to measure its age.

Whenever he could, Lyell, who had only entered the legal profession at the insistence of his father, dropped his law work to pursue his interest in geology. Twice a month he attended the Friday night meetings of London's Geological Society, and once he even travelled to Paris to learn at the feet of Cuvier himself. By 1826 he had become knowledgeable enough to write articles on geology for the respected *Quarterly Review*. When he discovered he could earn enough money from writing to finance geological expeditions, he abandoned his law career altogether.

In 1828, he began planning a journey to southern France. At this stage it hadn't entered his head that fossils could be used to measure geological time – that would happen during the journey. Instead he planned to settle a simmering controversy that had broken out several years earlier.

The dispute was precipitated – largely inadvertently – by the work of Cuvier. In 1805, while Napoleon's army invaded

A Fossil Clock

Charles Lyell.

Austria, and his navy fought the British at Trafalgar, Cuvier and his colleague Alexandre Brongniart had roamed the Paris basin, teasing out the secrets of the rocks. The whole basin around the French capital lay on a bed of chalk laid down at the end of the Secondary period (in modern terms, the end of the Mesozoic). However, the rocks Cuvier and Brongniart were interested in lay above this, and belonged to the Tertiary period. In geological terms, therefore, they were looking at comparatively recent events.

When Cuvier and Brongniart studied the succession of fossils within the strata, they noticed something strange. The animals that had lived in the basin had experienced several dramatic transitions. The sea-water creatures they found in one layer of strata had completely disappeared when they looked at the layer above. In their place they found land animals and freshwater creatures. In the layer above this, however, these in turn were replaced by sea-water creatures again.

Because the changes between one layer and the next were so sharply defined, Cuvier concluded they had been caused by swift and violent catastrophes – sudden moments of revolutionary upheaval, during which, in some parts of the globe, the bottom of the sea had lifted up to form land, while in others the land had plunged beneath the waves to form the sea-bed. It was these 'revolutions', Cuvier concluded, that had caused the changes of species.

The signs of the last of these catastrophes could still be seen in the uppermost layer of the Paris basin. Heaps of silt, sand and pebbles mixed with the bones of numerous animals bore witness to a terrible inundation. Perhaps, Cuvier speculated, this event, dimly remembered, accounted for the flood legends of ancient civilisations. Maybe, he suggested, the biblical Flood and the inundation that had caused the most recent upheaval in the Paris basin were one and the same.

When news of Cuvier's evidence for a flood reached England, it was greeted with delight by a small group of geologists who still hoped geology would verify the biblical account. At Oxford University the Rev. William Buckland, the newly appointed reader in mineralogy, was quick to take up the theme.

A colourful eccentric, Buckland approached geology with a chaotic enthusiasm. Bones, skins, skulls, stones, all lay scattered about his rooms. They even spilled over onto his breakfast table, where it was said that toast and trilobites fought for space. To add to the effect, he combined this love of chaos with an adventurous – some would say bizarre – culinary taste. Delicacies such as hedgehog, crocodile or bear were served to unwary visitors, while those in the know made their excuses. 'I have always regretted a day of unlucky engagement,' wrote John Ruskin, with heavy irony, 'on which I missed a delicate toast of mice.' Charles Darwin was not impressed. 'Though very good-humoured and good-natured [he] seemed to me a

A Fossil Clock

William Buckland.

vulgar and almost coarse man. He was incited more by a craving for notoriety, which sometimes made him act like a buffoon, than by a love of science.'

Buckland soon found evidence that persuaded him that Cuvier was right, and that Europe had recently been submerged beneath a tremendous flood. From the clay around London, he unearthed quartz pebbles that had originated in a rock formation at Leakey Hill, on the outskirts of Birmingham, over a hundred miles away. The stones, he decided, were too large for a river to have carried them such a great distance, so he concluded they had been swept there by a flood.

On the strength of this evidence, Buckland believed he had confirmed the events of Genesis. 'The grand fact of an universal deluge', he announced, 'is proved on grounds so decisive and incontrovertible, that had we never heard of such an event from Scripture or any other Authority, Geology of itself must have called in the assistance of some such catastrophe.'

Other geologists came to the same conclusion. The evidence was impressive. Deep parallel grooves scored into the sides of valleys, as if gouged out by flowing debris, giant boulders – known as erratics – scattered miles from their source, and large expanses of gravel strewn across southern England, all implied that in the not too distant past a violent flood, or at least a tumultuous wave, had hit Europe. What else but a flood would be powerful enough to carve the great river valleys of Europe? Even Sir James Hall in Scotland, one of Hutton's loyal supporters, was won over. Suspecting that a sudden uplift at the bottom of the ocean had created a tumultuous wave, he began setting off explosions of gunpowder under water to see if he could reproduce the tsunami in miniature.

However, not everyone was convinced. Several geologists thought Buckland had twisted the evidence to fit the Bible; above all in his claim that the Flood was universal. Many of the phenomena he described, such as the erratic boulders and diluvial deposits, occurred only in northern latitudes. If the Flood was worldwide – as he claimed – then why weren't the same phenomena found further south?

Chief among this group was Lyell. While studying at Oxford he had attended Buckland's lectures for three years in a row, and although initially inspired by his teaching, his views had subsequently moved away from his former tutor's. Above all, Lyell was horrified at Buckland's attempt to revive natural theology – the idea that nature could prove the truth of Scripture. He felt that the Bible and geology should be completely independent of each other, and that Buckland's statements could only damage the credibility of the fledgling science. In 1828 he set out for France in search of the evidence that would prove Buckland wrong, and quash once and for all the belief in the biblical Flood.

The trip was no shot in the dark. On an earlier visit to Paris he had learnt from the French geologist Constant Prévost that the alternate marine and freshwater strata of the Paris basin, which Cuvier had ascribed to catastrophic events, probably had a more prosaic explanation. Prévost had explained that the basin could have switched from freshwater lake to ocean bay and back again through small geological changes; a slight lowering of the barrier separating the basin from the ocean would let in the sea, while a tiny rise in the land would return it to a lake. There was no need to call upon catastrophes.

Even more compelling, and the reason for his journey, was the evidence of an English geologist named George Scrope. A few years earlier, Scrope had made a careful study of the small extinct volcanoes, known as *puys*, that lay scattered in the Auvergne region of south-central France. His findings implied that the landscape, far from being formed by a deluge, had in fact been created, step by step, by forces still at work. Lyell was captivated; Scrope's ideas were similar to his own. He was eager to see the evidence for himself.

He arrived in Paris at the beginning of May, and there met the two friends who would accompany him on the trip south: a young Scottish geologist named Roderick Murchison and his wife, Charlotte. Early on a Sunday morning, the party set out in Murchison's light open carriage for the Auvergne. Lyell was delighted to be with old friends. Mrs Murchison was 'very diligent, sketching, labelling specimens, and making out shells, in which last she is an invaluable assistant', while her husband was 'a famous hand at a bargain'. French innkeepers were notorious for doubling the price of a room at the first hint of an English accent, but with Murchison negotiating, Lyell noted appreciatively that 'we get off within a third of what the natives pay'. After six days on the road, they arrived in Clermont-Ferrand, the busy town in the centre of the Auvergne that would be their base for the next two weeks.

'Auvergne is beautiful,' wrote Lyell, 'rich wooded plains, picturesque towns, and the outline of the volcanic chain unlike any other I ever saw . . . There are innumerable old ruins for sketches, with lakes, cascades, and different kinds of wood, so that we wonder more and more that the English have not found it out.' There was no time for sightseeing, however; they had a job to do. Up at the crack of dawn, they began work at six and kept going through the midday heat, searching for the evidence that would prove that valleys weren't caused by the Flood.

Near the village of Pontgibaud on the river Sioule, they found one of the sites that Scrope had mentioned in his book. Standing on a ridge overlooking the valley, Lyell rapidly unravelled its history. Many years earlier, an eruption from one of the nearby volcanoes had sent a stream of molten lava into the valley below, where it had blocked the river's course. Since then, however, the river had slowly carved its way through the lava until it formed a new valley. This newly carved second valley provided a solid refutation of Buckland's claim that valleys were formed by the Flood. As Scrope pointed out, 'if the first excavation of these valleys is to be accounted for by . . . a deluge, to what are we to attribute the second . . . ? Not, most certainly, to a *second deluge*; for the undisturbed condition of the volcanic cones, consisting of loose scoriae and ashes, which actually let the foot sink ankle-deep in them, forbids the possibility of any great wave or debacle to have swept over the country since the production of those cones.'

Further on they came across a place where the river had cut its way past a lava flow creating a canyon a hundred and fifty feet deep. 'This is an astonishing proof of what a river can do in some thousand or hundred thousand years by its continual wearing,' remarked Lyell. Once again he observed that the crater and ash of the volcano were still intact, showing that no deluge had washed over the site since the lava had flowed into the valley thousands of years before.

A Fossil Clock

Lyell was lost in wonder at the immensity of time nature must have needed – not only to form this canyon, but all the other valleys of the world. What innumerable showers of rain it must have taken, to carve the valley of the Rhine. Scrope had described it perfectly: 'The sound which, to the student of Nature, seems continually echoed from every part of her works, is – Time! – Time! – Time!' That evening, Lyell returned to Pontgibaud knowing he had Buckland on the run.

From the Auvergne, Lyell and the Murchisons continued south towards Italy, collecting more evidence along the way. By now, Lyell was convinced that all geological features were the result of forces still at work today – rain, rivers, tides, earthquakes, volcanoes. Nothing else was required; no universal floods, no catastrophes, no revolutions, just these well-known, commonplace phenomena – and time. But how much time?

It was in this latter stage of his journey that he began formulating his idea for a fossil clock. The first clue came at the beginning of August, when Murchison fell ill with malarial fever, and the group stopped in Nice for a few weeks to allow him to recover. To pass the time, Lyell paid a visit to a local naturalist, Giovanni Risso, a middle-aged professor at the *lycée*, who had painstakingly built up a magnificent collection of fossil shells. When Risso proudly opened his cabinet for his inspection, Lyell was impressed. Inside were over two hundred different species, perfectly preserved, all drawn from a nearby Tertiary formation. Only 18 per cent, Lyell noted with interest, were fossils of species still living.

It was only a simple statistic, but it raised an important question. When had the other 82 per cent become extinct? Since Lyell had already dismissed the notion that the earth's history had been punctuated by catastrophes, he could not, like Cuvier, ascribe their disappearance to a sudden disaster. It therefore seemed probable that, in the same way that valleys had been eroded gradually, species had died out slowly, one at a time, over thousands if not millions of years.

Lyell decided to extend his trip. It would mean delaying his longed-for confrontation with Buckland, but that could wait, he was developing grander ideas. He was beginning to form a notion of how he could put a timescale on the strata of the rocks. If species had died out at a uniform rate, and new ones had been created at the same rate, then by comparing the ratio of extinct to living species in a given layer of rock it should be possible to form some estimate of its age. A rock formation containing half as many fossils of present-day species as another formation would be twice as old.

At Padua, the group parted company; the Murchisons returned to London while Lyell travelled on alone to Naples. He was heading for Sicily. Only there, he believed, could he find the answers to the 'thousand queries' he had raised in the Auvergne. Only on Etna could he hope to discover the timescale of the earth's past.

The inns had not improved since Brydone's day, but that didn't worry Lyell. He had come prepared, with plenty of tea and four bottles of brandy – supposedly to ward off malaria. Intent on his work, he combed the island, piecing together the order of the strata and the fossils they contained. Just as he expected, he found the shells of marine molluscs hundreds of feet up the mountain, evidence that the volcano had risen out of the sea. Descending the mountain, he searched further afield, looking for strata that had existed before the volcano started to grow. The lava fields stretched so far from the cone that it wasn't until he reached Syracuse, fifty miles away, that he found what he was looking for: a layer of blue clay that appeared to pass beneath the mountain itself. It contained a mass of well-preserved fossil shells, exactly what he needed to form some idea of the mountain's age. When he examined them, he found they were younger than he expected. In contrast to the shells he had seen in Risso's cabinet in Nice, of which only 18 per cent were of species still living in the Mediterranean, here 95 per cent belonged to present-day species. According to this 'fossil chronometer', the layer of clay that passed under Etna was barely old at all.

It was a revelation. 'I got so astounded by the results I was coming to,' he wrote to Murchison, '. . . that I began to doubt them.' He struck back across the island checking the strata, making sure of his facts. But there was no shaking the evidence. The mountain was born yesterday compared with the shells he had seen in Nice.

However, his chronometer was little use without some idea of its scale. For instance, how long did it take for 100 per cent of fossil species to die out and be replaced with new ones – what was the time span necessary for a complete revolution in the ocean fauna? To answer this question, Lyell attempted to discover how long it had taken for the last 5 per cent of fossils to die out; in other words, how long it had taken for Etna to grow above the layer of blue clay in which the fossils were found.

In an elegant piece of lateral thinking, he regarded the problem of dating Etna as similar to that of dating a tree. The succession of lava flows that had built up around the mountain over the years could be compared to a tree's growth rings. Obviously he couldn't cut through the mountain and count the layers, but he could make an estimate of the volume of lava contained in the mountain, then compare that with the rate at which Etna was known to erupt. It was only a rough estimate, but it was a start. Knowing that Etna was ninety miles in circumference, he noted that it would require ninety flows of lava, each a mile wide, to raise the foot of the mountain by the thickness of a single flow. Yet in two thousand years of human history there was no evidence that the altitude of Etna had varied at all. Most of the extra volume had come from the eighty or so minor cones scattered around the flanks of the mountain. Lyell estimated that, in 12,000 years, eruptions from these had added 'perhaps, several miles' to the mountain's diameter. After making a 'moderate computation' he concluded that the mountain was between 70,000 and 100,000 years old. In this way, he was able to put the rate at which fossil shells died out into perspective. A complete change in species, as had occurred in the Tertiary strata, would have taken in the order of a million or so years.

At the end of 1829, his work complete, Lyell headed for home, keen to do battle with Buckland and his supporters. As he passed through Rome, he learnt that a book on geology was about to be published in Dublin that aimed 'to Prove the Hebrew cosmogony'. His barrister's hackles rose. 'So much the better,' he replied. 'I have a got a rod for the fanatics, from a quarter where they expect it not.'

By the time Lyell arrived in London, Murchison had already struck the first blow. At a meeting of the Geological Society in December, he had read their joint paper on the formation of the valleys in the Auvergne. The seventy-strong audience had listened with more than usual attention as Murchison explained that the valleys were not created by a universal flood, but by rivers wearing away rock over a long period of time. After he sat down an angry debate had broken out, with Buckland furious, not only at his former pupil's betrayal, but that so many eminent members of the society were now siding with him.

By June 1829, with Lyell back in attendance, the diluvialists (supporters of the idea that the land had been shaped by giant floods) were on the run. In an attempt to refute Lyell's conclusions, one of Buckland's supporters, William Conybeare, read a paper purporting to show that 'no river, within times of history, has deepened its channel one foot!' It was a disaster. After describing all rivers as impotent, and dismissing the 'sluggish' Thames as 'scarce able to move a pin's head', let alone carve a valley, Conybeare sat down to listen to the following paper. By coincidence, it happened to be a surveyor's report describing how the force of a Cheviot stream had swept away a bridge. The juxtaposition of the two papers reduced Buckland and his supporters to a laughing-stock. When his desperate explanations became even more confused, a wit summed up his position in verse:

> Some doubts were once expressed about the Flood.
> Buckland arose, and all was clear as mud.

To be fair to Buckland though, much of the evidence he cited could not be explained by the steady-state processes Lyell proposed. It was only in the late 1830s that the real cause of his 'deluge' phenomena became apparent. Around this time, the Swiss geologist Louis Agassiz showed that glaciers not only moved, but were powerful enough to carve their way through solid rock. Geologists soon saw that glaciers had carved most of the geological features of northern Europe, from U-shaped valleys to erratic boulders and diluvial gravels. Buckland had been right to assume that powerful forces had been at work; however, what he had assumed to be the Flood had in fact been the Ice Age. Contrary to his image of a biblical 'stick-in-the-mud', Buckland abandoned his previous beliefs and became one of the first to accept and promote the new glacial theory. With his retreat, geology was finally wrested free of the Bible.

In 1830 Lyell published his ideas in a book, *Principles of Geology: Being an Attempt to Explain the Former Changes of the Earth's Surface by Causes now in Operation*. At the book's heart was his theory of uniformitarianism: the idea that throughout the history of the earth, the same forces that are acting today have always acted, and have done so at more or less the same rate. It had taken untold numbers of earthquakes to raise the mountains, step by step, until they had reached their present height; while the valleys had gradually been worn away, either by ocean currents while the land was beneath the sea, or, later, by the gentle erosion of streams and rivers.

Although this principle of gradual change formed the theoretical basis for his dream of a fossil clock, Lyell was eventually forced to concede that his attempts to devise a fossil chronometer offered no prospect of determining the earth's age. The lack of fossils in the deepest rock formations, he concluded, was not due to the fact that there had been no life on earth when they had formed, but was the result of intense heat and pressure deep within the earth destroying all traces of the life-forms that had existed. With no means of ascertaining how many countless generations of species had disappeared from the fossil record, he was forced to conclude, like Hutton, that the beginning of time lay 'beyond the reach of mortal ken'.

By 1840, though, the evidence that the earth was exceedingly ancient was overwhelming. After four decades of restless activity, geologists had pieced together nearly all the major geological classifications. The visible presence of countless generations of ancient species that had roamed the earth in former times left no doubt that the world had existed for far, far longer than the span indicated by the Bible. Despite this, Ussher's timescale refused to die. It lived on into the late nineteenth century, supported by theologians who found ways to reconcile this evidence with Scripture. Attempts at compromise were nothing new; from Burnet to Buckland, geologists had twisted their view of nature to make it agree with the biblical account. Now, however, the boot was on the other foot. Instead of geology being made to fit the Bible, the Bible had to be reinterpreted in the light of geology.

A Fossil Clock

The majority of churchmen – in Britain at least – supported one of two main interpretations. The first was embraced by the more orthodox Christians who stuck rigidly to the traditional notion that the Bible should be taken literally; that is, the world *was* created in six days. The interpretation that reconciled this dogmatic stand with the long timescale of geology was devised by a Scottish cleric named Thomas Chalmers, who proposed what later came to be known as the 'restitution' theory. This suggested that God had built the present world out of an old one. It was second-hand. According to Chalmers, the first verse of Genesis – 'In the beginning God created the heaven and the earth' – described the creation of the original world. However, between this verse and the next – 'And the earth was without form, and void' – an immeasurable period of time had elapsed. This was the version favoured by Buckland, who explained that 'millions of millions of years may have occupied the indefinite interval, between the beginning in which God created the heaven and the earth, and the evening or commencement of the first day of the Mosaic narrative.' According to this interpretation, it was during this 'indefinite interval' that the extinct species had roamed the earth. The description of the six days of Creation in the following verses of Genesis described the building of the second world – our own. Ussher's chronology applied solely to this later world.

The second solution, adopted by the more liberal clergy, followed the old notion that the six days of Creation were not literally days, but merely metaphorical periods of time – six eras that lasted an indefinite period. This idea was boosted by the confirmation that human beings were among the last species to appear on earth, just as the Bible said: God created them last. In this scheme, therefore, Ussher's chronology no longer applied to the whole history of the world, but only to the history of mankind. The planet might be indefinitely old, but mankind had only walked its surface for a few thousand years.

Aeons

While Lyell's theory appeared to close for ever all hope of finding the actual age of the world, it nevertheless opened up new horizons for pushing back geological time. His idea that only present causes had acted in the past, and that they had always acted at the same rate, would soon give geologists a tool for probing earth history in a way never possible before. The first person to make use of this idea to estimate the scale of geological time was a young naturalist who, in January 1832, was sailing towards South America aboard HMS *Beagle*.

9. Burnt Fingers

In the summer of 1831, Charles Darwin was considering training for the priesthood when out of the blue his Cambridge tutor, John Henslow, suggested he take up the offer of a passage on the survey ship *Beagle*, where he would work as the ship's naturalist. Twenty-two years old and fresh out of university, Darwin was hardly qualified for the voyage ahead. Although his knowledge of botany was good, and he had amassed a fine collection of beetles, his longest field trip hitherto amounted to no more than a three-week ramble in North Wales. Now he was to sail around the world! As Darwin hurriedly prepared for his departure, Professor Henslow – who supported Cuvier's 'catastrophist' geology – advised him to buy the first volume of Lyell's *Principles*, to take with him on the journey, 'but on no account to accept the views' it contained.

Darwin could hardly ignore them, however. From the first port of call onwards, nearly everything he read in Lyell's book was confirmed by the landscape around him. By the time he reached Chile he was converted. Any remaining doubts were literally shaken off when he experienced an earthquake for himself. He was resting in a forest at the time, lying down, when he felt the first tremor. 'It came on suddenly,' he recalled later, 'and lasted two minutes, but the time appeared much longer.' Two weeks later, when the *Beagle* sailed into Concepción harbour, he saw just how powerful the quake had been.

Not a house remained standing, piles of debris filled the streets, and the once magnificent cathedral had collapsed into 'a grand pile of ruins'. But his geologist's eye also noticed the signs of an enormous uplift. Along the coast, eyewitnesses reported that the land around the Bay of Concepción had risen between two and three feet, while the *Beagle*'s captain, Robert Fitzroy, discovered that the nearby island of Santa Maria had been lifted even higher. In one place he found beds of putrid mussels still clinging to rocks ten feet above the high-water mark. Here was evidence that present-day phenomena could raise mountains, step by step, just as Lyell predicted. When Darwin's account of this uplift was published some years later, it knocked on the head one of the last diluvialist notions: that the Andes had risen up in one sudden, cataclysmic upheaval that had launched a giant wave over the world.

Earlier in the journey, while anchored off Montevideo, Darwin had received the second volume of Lyell's book. It was quite unlike the first; Lyell had moved on from explaining rocks and strata to discuss the living world. Why had some animals become extinct? Where did new species come from? Difficult questions, that Lyell had struggled to answer.

Extinction, he believed, was the inevitable result of a changing climate. If the world was passing through a period of falling temperatures (as Lyell believed it was), then animals would be forced to migrate from their usual territories to seek warmer habitats, either by descending to lower altitudes or by moving nearer the Equator. As these creatures entered a new region they would cause the extinction of the previous occupants, left 'sickly and almost incapable of defence' by the change in temperature.

New species were harder to explain. Lyell, a practising Christian, thought that each species had originated from a single pair of individuals – rather as mankind had supposedly originated from Adam and Eve. These pairs were created at a time and place that allowed them to thrive. He never explained how they appeared or where they came from; they just cropped up from time to time in the fossil record, created by some unknown process.

Lyell's book drew Darwin's attention to one of the central questions in science, 'that mystery of mysteries' as the astronomer Sir John Herschel called it: 'the replacement of extinct species by others'. Herschel was the leading British scientist of the age. In June 1836 he was living in South Africa, mapping the constellations of the southern sky, when the *Beagle* anchored off the Cape. Three days later Charles Darwin bounded up the steps of his house, eager to talk. For years he had admired Herschel's work; it was unthinkable to pass through Cape Town without paying him a visit.

As it happened, Herschel had also read Lyell's book and been similarly inspired. Enthused by this common interest, the two men strolled through Herschel's 'charming' garden discussing volcanic eruptions and continental upheaval. No one knows how far their conversation ranged that day, but Herschel had already identified the creation of species as the key unanswered question in Lyell's book. Only four months earlier he had written to Lyell in London praising him for raising the subject, and exhorting his fellow scientists in verse to seek an explanation for the appearance of new species:

> He that on such quest would go must know nor fear nor failing
> To coward soul or faithless heart the search were unavailing.

It was a wide-ranging letter, part scientific explication, part passionate call to arms. Most impressively, Herschel had picked up Lyell's uniformitarian idea and extended it to linguistics, wielding it with consummate skill to pierce the last bastion of biblical chronology: the age of man.

In an attempt to gauge the scale of human history, Herschel called for an inquiry into the 'laws of verbal corruptions', and the 'process and rate' at which words changed their meaning. Just as landscapes changed over time, so did languages, he argued, and their rate of change could be used to determine how long man had existed on this earth. 'Words are to the Anthropologist what rolled pebbles are to the Geologist,' he wrote,

– Battered relics of past ages often containing within them indelible records capable of intelligible interpretation – and when we see what amount of change 2000 years has been able to produce in the languages of Greece & Italy or 1000 in those of Germany, France & Spain we naturally begin to ask how long a period must have lapsed since the Chinese, the Hebrew, the Delaware & the Malesass [the language of Madagascar] had a point in common with the German & Italian & each other. – Time! Time! Time! – we must not impugn the Scripture Chronology, but we *must* interpret it in accordance with *whatever* shall appear on fair enquiry to be the *truth* for there cannot be two truths. And really there is scope enough: for the lives of the Patriarchs may as reasonably be extended to 5000 or 50,000 years apiece as the days of Creation to as many thousand millions of years.

This was fiery stuff; even Darwin was startled. If he didn't hear Herschel's views for himself in Cape Town, he certainly read about them four months later on his return to England. 'As far as I know everyone has yet thought that the six thousand odd years has been the right period,' he wrote to his sister, 'but Sir J. thinks that a far greater number must have passed.' Darwin would soon think the same. The last stronghold of Ussher's chronology – the span of man's existence – was about to fall.

Back in England, Darwin settled in London, not far from the home of Charles Lyell, who quickly became one of his closest friends. His letters from the *Beagle* had already won him fame, and with Lyell's encouragement he was elected to the Geological Society. Within a year he had begun a notebook on the origin of species. Although a passionate supporter of Lyell's uniformitarian geology, Darwin disagreed with his views on creation. He sided with Herschel in believing there had to be some underlying law of nature that determined the creation of new species. 'We can allow satellites, planets, suns, universe, nay whole systems of universe to be governed by laws,' he lamented, 'but the smallest insect, we want to be created at once by special act.' Lyell's assertion, that species once created never altered, flew in the face of the evidence. Present-day South American animals bore an uncanny resemblance to extinct species: the living armadillos, for instance, were much smaller than their fossil counterparts, yet in almost all other respects they appeared identical. To Darwin, such slight variation implied the old species had transmuted into the new – they had evolved.

This idea was not new. Over thirty years earlier, the French naturalist Jean-Baptiste Lamarck had suggested that all the present species had developed from earlier, simple forms of life. The question Darwin now attempted to answer was: 'How?'

For eighteen months he toyed with various ideas, filling pages of his notebook with jottings and diagrams, searching for the mechanism which drove evolution. Sex was important, he could see that. Combining traits from two individuals created variety, which was a starting point for evolution. But it didn't seem to go anywhere; this variety was directionless. For however many giraffes were born with long necks there would be an equal number born with short necks. As the population interbred these characteristics would mix together again, preventing any long-term change. It appeared a dead end. Only if there existed some mechanism that favoured one characteristic over another could new species evolve.

For some time, the key to transmutation continued to elude him. Then, in October 1838, he 'happened to read for amusement' Thomas Malthus's *Essay on the Principle of Population*, first published forty years earlier. Malthus was a clergyman and political economist who railed against social welfare and hand-outs to the poor. He also argued against charity, claiming it only led to an increase in population, which ultimately meant more starving mouths to feed from the same limited resources. It was a cold philosophy, one with which Darwin was already familiar, but what fascinated him were Malthus's statistics. They showed that if the human population was allowed to grow unchecked it would double every twenty-five years; within a short time the planet would be overrun.

Darwin realised this didn't happen; the population didn't increase because there was already a struggle for existence. Disease, famine and warfare killed thousands, if not millions of people each year. The same applied to animals. Life was a competition to survive. As the bitter struggle unfolded, animals endowed with favourable characteristics – ones that helped them to survive – would be more likely to pass these on to future generations, while unfavourable characteristics would die out. This was *natural selection*. Over successive generations, the gradual accumulation of favourable characteristics would amount to significant change; given enough time, a new species would evolve.

For the next twenty years Darwin sat on his idea, keeping it secret from all but a handful of his most trusted friends. For the sake of his scientific credibility he wanted to be sure of his facts, but he also knew that his ideas were political

Burnt Fingers

Charles Darwin, in a photograph taken around 1854, shortly before he published *The Origin of Species*.

dynamite. Socialist reformers had taken to championing versions of Lamarck's evolutionary theory as a rallying cry for the working classes. If species could progress and develop, they argued, then so could the downtrodden masses. The idea of evolution came to be associated with socialist reform and atheistic revolt: the church condemned it as morally degenerate; the political establishment regarded it as dangerous. Before announcing his ideas to the world, Darwin had to be sure he was right.

Through two decades, while he anguished over his theory, life rolled by. He married his cousin, Emma Wedgwood, moved to a large house in the country, and fathered ten children, seven of whom survived to adulthood. Throughout this time he was tortured by recurring illness. Boils, vomiting, flatulence, hysterical crying, shivering, and dying sensations, punctuated his daily routine. No one knew what caused these outbreaks, but the stress of his unorthodox, secret work seemed only to make them worse. He tried everything to find a cure: dieting, sucking lemons, sipping acids, ice-cold baths, cold-water douches. And when these failed he resorted to more technical quackery: galvanised plates on his stomach, and 'electric' chains around his neck. Nothing worked.

On the days he was physically capable, he continued to work on his theory. He accumulated evidence, checked facts, and investigated every objection to his hypothesis he could imagine. Of all the possible objections, time appeared the least significant; yet it was essential to his theory. Natural selection worked slowly, its small changes taking thousands of years to produce significant variations. To affect the astonishing changes from sea creatures to mammals would have required many millions of centuries. Thanks to Lyell, however, Darwin believed time was available in unlimited quantities. It barely crossed his mind that it might be a problem.

By the mid-nineteenth century, the British public was beginning to grasp the enormity of the earth's history. The age of mankind was still popularly believed to be about 6,000 years, but prehistoric time had expanded beyond comprehension. By 1850 geologists had identified and named nearly all the sedimentary strata. Words such as Oolite, Devonian and Silurian littered the pages of up-market periodicals, while readers of popular magazines marvelled at illustrations of the strange creatures that had inhabited this ancient world. In 1854, Londoners flocked in their hundreds of thousands to the south London suburb of Sydenham to visit the re-erected Crystal Palace, and gaze in wonder at the life-sized concrete dinosaurs constructed in the nearby park. Two huge Iguanodons, a Pterodactyl and a fearsome Megalosaurus dominated an island in the middle of an ornamental lake, while other wondrous creatures lurked at the water's edge. The designer had arranged them in chronological order to enable the Victorian public to imagine the endless eras in which they had lived. But old notions died hard.

'What are those?' gasped a startled traveller when he saw the creatures from the window of a passing train.

'Antediluvian monsters,' replied a fellow passenger.

'Why antediluvian?' asked the first.

'Because they were too large to go into the ark; and so they were all drowned,' came the reply. Geologists may have dismissed all notion of the Flood, but the public perception of history was still shaped by the Bible.

Twelve miles away, in his country home at Downe, Darwin laboured on, absorbed by a thousand questions on the variability of species. He was so confident that he had 'almost unlimited time' for natural selection to do its work that by 1858, a full twenty years after he first thought of his theory, he had made only a cursory examination of the question. It was too little, too late. In June that year, an envelope arrived from the other side of the world.

It brought him up short. What he had secretly dreaded for years had finally happened. Alfred Russel Wallace, an English naturalist-collector working in the Malay Archipelago, had hit upon the same idea. 'I never saw a more striking coincidence,' Darwin groaned to Lyell, 'if Wallace had my M.S. [manuscript] sketch written out in 1842 he could not have made a better short abstract! Even his terms now stand as Heads of my Chapters.' With the theory out in the open, the pressure was on for Darwin to publish.

Pushed for time, but needing to convince his readers of the earth's antiquity, Darwin wrote to Andrew Ramsay, a geologist working for the Geological Survey in Scotland, requesting information on the maximum thickness of the sedimentary rocks in the British Isles. Even though Darwin believed there were huge gaps in the geological record, he knew that nothing would convey the enormity of prehistoric time so well as the vast thickness of rock built up slowly from sediment deposited on the ocean floor.

From the figures he obtained from Ramsay and others, Darwin added up the thicknesses of the various formations to produce depths for each of the geological eras:

	Feet
Tertiary strata	2,240
Secondary strata	13,190
Palaeozoic strata (not including igneous beds)	57,154

Altogether, they came to 72,584 feet, or nearly thirteen and three-quarter miles. An astonishing figure, that still conveyed only 'an inadequate idea' of the time that had elapsed.

To try to put this into perspective, Darwin made a hasty attempt to find the age of the Weald, the wide valley that stretched between the North and South Downs of south-east England. The valley had been carved through chalk, the last rock to be deposited in the Secondary era. By finding out how long it had taken to erode, Darwin could obtain a rough estimate for the duration of the most recent geological era, the Tertiary period.

He knew the Weald well – it lay barely five miles from his home – and he was familiar with the standard explanation of how it had formed. According to Lyell, who had devoted two chapters of *The Principles* to the subject, the Weald had once been a narrow channel flanked by chalk cliffs. Over innumerable centuries the sea had raced through this gap, progressively wearing back the cliffs until the channel had reached the size of the present-day valley, at which point a long and continuous upheaval had raised it from the water.

On the back of a single piece of paper, Darwin dashed off a rough estimate for how long it had taken the sea to wear

Darwin's preliminary calculation for the age of the Weald (bottom left) made in 1857. The discrepancy with the published figure is due to an error in his arithmetic that he corrected the following year.

away the cliffs. He knew the distance from the North to the South Downs was twenty-two miles, and the thickness of the formations was, on average, 1,100 feet, but he didn't know how fast the sea eroded chalk. In a preliminary calculation made two years earlier he had guessed that '1 inch in 100 years' was about the right figure for a cliff 500 feet high. The deeper valley of the Weald, he assumed, would have taken proportionately longer to erode. Using the above figures, Darwin calculated that 'the denudation of the Weald must have required 306,662,400 years; or say three hundred million years.'

This impressive figure gave only the duration of the Tertiary period – a tiny fraction of geological time. When the whole depth of strata was considered, the amount of time available for natural selection became so vast as to be almost unimaginable. 'What an infinite number of generations which the mind cannot grasp, must have succeeded each other in the long roll of years!' he wrote, confident that his theory had all the time it needed.

When Darwin inserted the result of his Weald calculation into *The Origin of Species* in 1859, he had no idea how bold a claim he was making. 'In all probability,' he asserted unhesitatingly, 'a far longer period than 300 million years has elapsed.' It was only that winter, when the reviews began to arrive, that he realised he had made a ghastly mistake.

At first his critics were too busy venting their spleen on the idea that man had descended from an ape to even notice the Wealden calculation. But by December an astute critic had spotted the errors, and on Christmas Eve *The Saturday Review* gave Darwin a damning write-up. The anonymous author pulled no punches. Focusing on Darwin's estimate of geological time, he accused him of exaggerating the age of the earth and bungling his calculation of the age of the Weald. His mistakes were as plain as a pikestaff. First he hadn't allowed for the fact that the sea would eat away the cliffs on *both* sides of the channel at the same time, which would halve the time it took to erode. Second, he had grossly underestimated the speed of erosion; the waves only needed to erode the foot of a cliff to bring the whole edifice tumbling down, therefore a cliff 1,100 feet high would crumble away just as fast as one 500 feet high. Finally, the sea had probably not worn away the chalk, but the softer greensand that underlay it, which would also speed up the whole process. In short, Darwin had 'committed some manifest errors': he had 'enormously overrated' the amount of time available, and had based his 'speculations' on 'a pile of unsupported conjecture'.

Darwin realised he had blundered. 'Some of the remarks about the lapse of years are very good,' he admitted to Lyell, 'and the reviewer gives me some good and well deserved raps – confound it, I am sorry to confess the truth.' Forced to retreat, he moved quickly to limit the damage; in January he sent the publisher of the American edition a last-minute foot-note admitting he had erred.

I beg the reader to observe that I have expressly stated that we cannot know at what rate the sea wears away a line of cliff: I assumed the one inch per century in order to gain some crude idea of the lapse of years; but I always supposed that the reader would double or quadruple or increase in any proportion which seemed to him fair the probable rate of denudation per century. But I own that I have been rash and unguarded in the calculation.

But with attention now focused on the Wealden calculation, the critics closed in for the kill. In February 1860 he took a crushing blow when no less a person than John Phillips, that year's president of the Geological Society, slated the calcu-lation in his presidential address. Phillips, who had succeeded

Buckland as professor of geology at Oxford University, took violent exception to the theory of natural selection. He saw the Wealden calculation as Darwin's Achilles heel and used his address to castigate him for his 'abuse of arithmetic'. To reinforce his argument, Phillips then decided to make his own estimate for the age of the earth's crust.

His approach was inspired. Instead of calculating how long it took to erode a geological formation, as Darwin had done, he chose to calculate how long it took to build one up. Phillips realised that if he could find the thickness of the layer of sediment brought down by the rivers of the world and deposited on the ocean floor each year, then – assuming it had always been deposited at the same rate – he could calculate how many years it would have taken for the sedimentary rocks to reach their present depth.

Searching through old scientific journals, Phillips found a report of an experiment that had been carried out in India thirty years earlier. During the latter half of 1831, a Rev. Robert Everest had measured the quantity of 'earthy matter' carried by the Ganges as it flowed through Ghazipur. He had scooped water from the centre of the river, evaporated it and weighed the solid remains to produce figures for the volume of material carried in the hot, rainy, and winter seasons. Adding them together gave the annual discharge of sediment from the Ganges: a massive 6 billion cubic feet. Armed with Everest's figures, Phillips worked out that if this dispersed over the ocean floor to cover an area equivalent to the Ganges' catchment area – roughly 300,000 square miles – then each year the mud flowing from the mouth of the river would add another $\frac{1}{111}$ inch to the thickness of the strata. At this rate, taking Darwin's figure of 13¾ miles for the total thickness of all the sedimentary strata, Phillips calculated that the first layer had been laid down 96 million years earlier.

Of course the calculation involved several large assumptions, and Phillips didn't for a moment believe this figure was the true age of the world. But he was certain his argument was strong enough to quash Darwin's date once and for all.

Sure enough, Darwin was crumbling. Although he later wrote that he 'would rather have been well attacked than have been handled in the namby-pamby, old-woman style of the cautious Oxford Professor', Phillips' criticism hurt. He wrote to Lyell, who was preparing a book on the antiquity of man, with a word of warning: 'Having burnt my own fingers so consumedly with the Wealden, I am fearful for you; but I know how infinitely more cautious, prudent & far-seeing you are than I am. – But for Heaven-sake take care of your fingers; to burn them severely, as I have done, is very unpleasant.'

The news wasn't all bad, however. Recent discoveries in Europe bolstered his evolutionary views. For the first time there was concrete evidence that man's history stretched back further than the six thousand years of the biblical account. Four years earlier, quarrymen in Germany had uncovered the first Neandertal skull from a cave near Düsseldorf; and only the previous year Lyell, after seeing ancient arrowheads in the flint beds of the river Somme in France, had declared his conviction that humans had been living and hunting since before the Ice Age. The idea of man's recent creation was no longer defensible.

Perhaps the most unexpected support, though, came from Leonard Horner, Lyell's ageing father-in-law, who in 1861 had succeeded Phillips as president of the Geological Society. Putting aside his geological interests, Horner had turned his attention to the Bible, and in particular Ussher's date for the creation of the world. He had become convinced that the main reason people still believed that man and the world were of recent origin was because they assumed that the date of 4004 BC was actually part of the Bible. By now hardly anyone remembered that it was the work of Ussher.

From inquiries at Oxford, Cambridge, Edinburgh, and the Queen's printers in London, Horner tried to find out who had authorised the insertion of the date; without success. A tour of old libraries, however, revealed that its inclusion was widespread and dated back a long time. From the musty shelves of the British Museum he unearthed a 1701 Bible which contained the date in its margin, while a visit to the Royal Library in Berlin revealed Ussher's date in a French translation from the following year.

In February 1861, instead of delivering the usual treatise on geology, he surprised the members of the Geological Society by devoting a large part of his presidential address to a critique of the Bible, and in particular, Ussher's marginal dates. 'It is more than probable,' he declared,

> that of the many millions of persons who read the English Bible, a very large proportion look with the same reverence upon that marginal note as they do upon the verse with which it is connected . . .
>
> The youth is told in the morning at school, probably by his own minister of religion, as I myself have witnessed, that not more than about 6000 years have elapsed since the creation of the world. In the evening he may attend a lecture on geology, very possibly by one of the ninety-three clergymen who are Fellows of this Society, and hear that, in a work just issued from the press, it is stated that 'the probable length of time required for the production of the strata of coal, sandstone, shale, and ironstone in South Wales is half a million of years.' It is thus easy to see what a confusion must be created in the youth's mind, and that he will involuntarily ask himself, 'Which of the two statements am I to believe?'

Burnt Fingers

Horner hammered home his message:

> To remove any inaccuracy in notes accompanying the authorized version of our Bible is surely an imperative duty. The retention of the marginal note in question is by no means a matter of indifference: it is untrue, and therefore it is mischievous.

When he heard about the lecture, Darwin was amazed. Despite having spent two years at Cambridge training for Holy Orders, he had no idea the date was Ussher's. 'How curious about the Bible! I declare I had fancied that the date was somehow in the Bible,' he wrote to Horner, adding admiringly: 'You are coming out in a new light as a Bible critic.'

Horner's plea, however, appears to have fallen on deaf ears. Publishers continued to print Ussher's date in the margins of bibles for many years to come.

Darwin's own date, however, lasted barely two more months. Swallowing his pride, the previous year he had been forced to admit that the 'confounded Wealden calculation' was no longer tenable, and had asked for it to be 'struck out' of future editions of his book. When the third edition of *The Origin of Species* appeared in April, all mention of the Weald had gone.

Relieved to be rid of 'those confounded millions of years', Darwin relaxed; he believed he had put the question of the earth's age behind him. And for the next six years it appeared that he had. But throughout this time a storm was brewing, and when it broke it threatened to sink his whole theory.

10. Doomsday Postponed

The man laid up in bed with a broken leg detested Darwinian natural selection. He disliked the smugness of Darwin's supporters, their apostolic airs, and the evangelical conviction with which they spread their heathen creed; he was irritated by their certainty that man had evolved by natural selection; but, more than all this, he fumed at their assertion that the world had existed indefinitely. In the opinion of the Scottish physicist William Thomson – the future Lord Kelvin – Darwin and his supporters were wrong. Evolution was an impossibility, and the lengthy timescale on which it relied, a castle in the air. Sure that so vast a span of time flouted the established laws of physics, Kelvin resolved to bring both it and natural selection crashing to the ground.

Given a choice, the last person Darwin would have clashed with was Kelvin. Lively, hard-working and stunningly intelligent, he towered over Victorian physics like a colossus. At the age of just twenty-two he had been appointed professor of natural philosophy at Glasgow University, and by twenty-eight had formulated the second law of thermodynamics – one of the cornerstones of modern scientific thought. Later, for his achievements as a physicist and inventor, he became the first British scientist to be raised to the peerage, and today his pioneering role in thermodynamics is commemorated throughout the world in the Kelvin scale of temperature. A regular churchgoer, who began his university lectures with prayers, Kelvin believed that God had created mankind, just as he had created the universe. Human beings were not the latest product of some cold, mechanistic process, but part of God's design. Spurred on by this hostility to Darwin's theory, he would find three independent ways of determining the age of the world. All three would indicate that the lifespan of the earth was too short for natural selection to have taken place, and all would concur so closely with each other that they would shake Darwin's confidence in his theory.

In 1860, the year after Darwin published *The Origin of Species*, Kelvin was still in his mid-thirties and leading a hectic life. In addition to his research and lectures at Glasgow, he commuted regularly to London, where, as a director of the Atlantic Telegraph Company, he was actively involved in preparations to lay a cable between Europe and America. Although he was aware of Darwin's theory from the rumpus it had caused at the British Association meeting that summer, nothing could have been further from his mind as he rushed to keep up with his busy schedule.

Then in December, while curling with some friends on a frozen pond, he slipped on the ice and fractured his leg. It was a bad break, at the neck of the left thighbone, and it forced him to hole up in the coastal town of Largs to recuperate. Stuck in bed through the long winter months, with little to do but watch the ships sailing up the Clyde to Glasgow, or listen to the waves breaking on the pebble beach outside his window, Kelvin's mind drifted back to Darwin's theory.

Long before the publication of *The Origin of Species* he had harboured the conviction that the premise on which it was based – uniformitarian geology – was unsound. The world had not existed indefinitely, as Lyell insisted, but had a finite and relatively short life. Kelvin knew this from his studies of the sun. Just like the fire burning in the hearth beside his bed, the sun's energy supply was limited. Like any other fire that is not replenished, the sun would eventually burn up its supply of fuel and fizzle out.

How long this would take was a tricky question. Years earlier, physicists had assumed that the sun was powered by some sort of chemical reaction, such as burning coal. However, when Kelvin had calculated how much energy a chemical fuel could supply, he had discovered that if the sun was powered this way, it would have burnt out in just 3,000 years. This clearly hadn't happened. But if the sun wasn't burning conventional fuel, where did its energy come from?

Kelvin decided that the answer lay in gravity. Borrowing an idea from the German physicist Hermann von Helmholtz, he proposed that the sun had initially been formed and fuelled by a large cloud of meteors. Pulled together by their mutual gravitation, the meteors converged into an increasingly tighter and tighter ball; as they did so, they collided so frequently, and with such force, that they generated a phenomenal amount of heat. When the temperature reached the melting point of rock, they liquefied into an incandescent molten mass. The sun was born. From then on, its heat was maintained by the energy released as the giant mass slowly contracted under gravity. According to Kelvin's calculation this would supply enough energy to last 20 million years. Far shorter than the time Darwin needed for natural selection to take place.

It was all the evidence Kelvin needed. In the spring of 1861, still hobbling around the house at Largs, he wrote to his friend, the physicist James Joule, in Birmingham, outlining his intention to attack Darwin's theory. Although his letter hasn't survived, Joule's reply urged him on.

I am glad you feel disposed to expose some of the rubbish which has been thrust upon the public lately. Not that Darwin is so much to blame because I believe he had no intention of publishing any finished theory but rather to indicate the difficulties to be solved . . . It appears that nowadays the public care for nothing unless it be of a startling nature. Nothing pleases them more than parsons who preach against the efficacy of prayer and philosophers who find a link between mankind and the monkey or gorilla.

At the same time, Kelvin also wrote to John Phillips at Oxford, seeking his opinion of Darwin's estimate. His reply couldn't have been more encouraging.

My Dear Professor, –
. . . Darwin's computations are something *absurd* as to the wasting of the sea. For the calculation includes as a coefficient the *height*! of the cliff. This astonishing error is not the only one. No one who ever does calculate (among our geologists) attaches any weight to the result! In my *Life on Earth* I have given some calculations which, for the period of stratification, rise to 96 millions only.

Given their different methods of tackling the problem, this wasn't so far from Kelvin's own figure of 20 million years for the age of the sun. Especially since Kelvin knew that small changes in his initial assumptions could significantly alter his result. When he allowed for these, he concluded that it was 'most probable that the sun has not illuminated the earth for 100,000,000 years, and almost certain that he has not done so for 500,000,000 years'.

Too ill to attend the Manchester meeting of the British Association that September, Kelvin arranged for a colleague to read his paper in his absence. Intending to damage Darwin's theory, he must have been sorely disappointed. If the geologists and Darwin's supporters noticed his paper at all, they chose to ignore it. Whether this was due to Kelvin's absence, or to the hesitant wording of his figures, is impossible to say; either way, the message fell on deaf ears.

Frustrated by the lack of reaction, a year later Kelvin launched a fresh assault. This time he chose to attack the geologists on their home ground – the age of the earth. Again, it was heat that fuelled his argument. Like Buffon, he was certain that the earth, in earlier times, had been a molten sphere, and that it had subsequently cooled to create the planet we see today. However, unlike Buffon – who had been forced to model the earth with miniature spheres – Kelvin was able to work with direct evidence. He knew that the temperature beneath the earth's surface increased by 1°F for every 50 feet travelled towards its centre. He also knew the conductivity and specific heat capacity of the most common rocks, and the temperature at which they melted. To calculate how long the earth had taken to cool to its present temperature, all he had to do was drop these figures into an equation the French mathematician Joseph Fourier had derived fifty years earlier.

'Fourier's great mathematical poem', as Kelvin called the equation, described how heat flows through solid materials. It is said that after formulating it, Fourier himself had calculated the age of the world, but was so staggered by the result that he immediately destroyed the calculation, and never told anyone of his discovery. Whatever the truth of this story, when Kelvin made the calculation in 1862 he found that the earth had solidified 98 million years earlier – a remarkably close fit to Phillips' figure of 96 million years for the age of the earth's crust.

When Kelvin published his results in the *Transactions of the Royal Society of Edinburgh* later that year, Darwin's supporters once again paid scant regard. It's possible that they didn't notice the conflict with their own timescale. By the time Kelvin had allowed for errors, his estimate for the age of the earth varied between 20 million and 400 million years. The result was neither short enough, nor certain enough, to upset Darwin.

Indeed, if anything in Kelvin's work worried Darwin at this time, it was not his estimate of the earth's past, but the alarming picture he painted of its future. Kelvin's vision was apocalyptic. With the sun running out of fuel, he explained, 'the inhabitants of the earth cannot continue to enjoy the light and heat essential to their life, for many million years longer, unless sources [of energy] now unknown to us are prepared in the great storehouse of creation.' In this bleak scenario doomsday was approaching, and it was approaching fast.

In his mind's eye Darwin had imagined a bright future for life on earth. Evolution would lead to a more civilised race; it would create a world free of slavery and inhumanity, a new golden age in which the planet would be populated with intelligent, caring people. Given time, 'mankind will progress to such a pitch,' he declared, that the present generation would be looked back on 'as mere Barbarians'. Kelvin's vision filled him with despair.

> Even personal annihilation sinks in my mind into insignificance compared with the idea, or rather I presume certainty of the sun some day cooling & we all freezing. To think of the progress of millions of years, with every continent swarming with good & enlightened men all ending in this; & with probably no fresh start until this our own planetary system has been again converted into red-hot gas – sic transit gloria mundi [So passes away the glory of the world], with a vengeance.

Although Kelvin's attempts at whipping up a storm of opposition to Darwin's theory had so far failed, a few years later he pitched in again with a new argument, more subtle and

ingenious than the previous two. This time, he based his calculation on the speed of the earth's rotation. Kelvin realised that
the spin of the earth on its axis was gradually being slowed
by the friction of the tides. The pull of the moon, and to a
lesser extent the sun, created two bulges of water on either
side of the earth that acted like brakes squeezing the spinning
planet between them, and ever so slightly slowing its rotation.
Like a top winding down, the earth spun slower each year.
His calculations were a little imprecise (the amount the earth
slows depends on the relative distribution of the land and the
sea); nevertheless, he concluded that the planet was slowing
down by a little over $\frac{1}{5}$ of a second per year, or 22 seconds
per century. In order to use this figure to determine its age,
however, he needed to know how fast it was spinning when
it first formed.

This wasn't as impossible as it seemed. Ever since Newton's
day, it had been widely believed that the original speed of the
earth was indelibly preserved in its shape: the faster the spin
when it solidified, the larger the bulge at the Equator. When
Kelvin investigated the problem, he concluded that the bulge
was as large as, but no larger than, it would be if the earth
had formed with a centrifugal force 3 per cent greater than
at present. Given its present rate of retardation, this suggested
that the earth had solidified 100 million years ago.

A consensus had emerged; Phillips' Ganges calculation and
all three of Kelvin's methods had placed the age of the world
in the region of 100 million years: much younger than the
time required by Darwin. In 1868, Kelvin abandoned his
softly-softly approach and attacked the geologists directly. In
a provocative address to the Glasgow Geological Society in
February that year, he threw down a challenge. 'A great reform
in geological speculation seems now to have become necessary,' he announced. 'It is quite certain that a great mistake
has been made – that British popular geology at the present

time is in direct opposition to the principles of natural philosophy.'

Kelvin could no longer be ignored. And not just because of the strength of his scientific arguments. In the preceding years he had become a national hero. The enterprise to lay a telegraph cable across the Atlantic had finally succeeded, and it was Kelvin's invention – a delicate receiver that could detect the faint electrical signals – that had made it all possible. With success came public recognition: the people hailed him as a genius, Queen Victoria gave him a knighthood, and, thrust into the limelight, it was inevitable that his dispute with Darwin and the geologists would spill over into the public arena. The gloves were off. The conflict was out in the open.

The dispute split the British scientific community into two camps. In the north lay Kelvin and the physicists, who believed that the earth had existed for 100 million years or less; in the south sat Darwin and the uniformitarian geologists, who believed it had existed indefinitely. Throughout 1869 furious articles flew from the presses as supporters of each camp attacked each other's position. From the start, it was apparent that Kelvin's camp had the upper hand. The loose speculations of geology were little match for the mathematical precision of physics. Darwin's 'bulldog', Thomas Huxley, put up a spirited defence at the Geological Society's annual meeting in February, but even he was forced to concede that Kelvin might be right, and that the world might only be 100 million years old.

By the autumn Darwin's nerves were frayed. 'I felt very small when I finished the article,' he conceded, after reading another blistering attack, but deep inside he still felt sure that the world would be found 'rather older than Kelvin makes it'.

By the following year, however, his support had crumbled. Even old allies such as Lyell and Wallace deserted him. Lyell now estimated that all the sedimentary strata had been laid down in a period of 240 million years; while Wallace – who had come to believe that drastic changes in climate had accelerated the rate of evolutionary change – now thought that 24 million years would suffice. Lacking the mathematical knowledge to question Kelvin's arguments, Darwin felt hopeless. When his son George, who was studying mathematics at Cambridge, arrived home that spring, Darwin quizzed him on Kelvin's calculations. The news was bad; George too sided with Kelvin.

Over the last twenty years of his life, Darwin revised *The Origin of Species* several times, toning down his initial claims for the earth's longevity. He quietly removed his more assertive phrases, such as 'infinite number of generations' and 'the long roll of years', at the same time acknowledging that Kelvin's arguments posed 'a formidable objection'. Some of the fighting spirit remained, however. He defiantly pointed out that Kelvin's wide margins of error showed how 'doubtful the data are', and hinted obscurely that 'other elements may have to be introduced into the problem', but in private the Scottish physicist continued to loom 'like an odious spectre' that would haunt him for the rest of his days.

Shortly before his father's death, George Darwin wrote to Kelvin summing up how his father's views of the age of the world had changed since the publication of the first edition of *The Origin*. 'I have no doubt that if my father had had to write down the period he assigned at that time, he would have written a 1 at the beginning of the line & filled the rest up with O's. – Now I believe that he cannot quite bring himself down to [the] period assigned by you but does not pretend to say how long may be required.' He never would. Having burnt his fingers with the Wealden calculation, he kept his thoughts to himself, and left pronouncements of the earth's age to a younger generation.

★   ★   ★

Doomsday Postponed

In 1882, the year that Darwin died, Kelvin's star was in the ascendant. The transatlantic telegraph cable had made him a wealthy man (at one pound per word the enterprise was practically a licence to print money), and from his share of the patent rights he had bought a 126-ton schooner, the *Lallah Rookh*, and built himself a stately home. His reputation was unassailable and his advice sought on all matters scientific, however bizarre. When Gilbert and Sullivan wanted the fairies in their opera *Iolanthe* to wear electric lights in their hair, they called on Kelvin.

There was one thing, however, that Kelvin couldn't fix – the disappearance of Ussher's dates from the Bible. Whatever the true age of the earth, it was obvious to just about everyone that it was far older than the dates set down in the margin of the Old Testament. In 1885, after fifteen years of committee meetings attended by the most learned theologians from both sides of the Atlantic, the Church of England finally approved the publication of a revised edition of the Authorised English Bible. When he opened his copy, Kelvin was disappointed to find that Ussher's dates had gone. His niece Agnes Gardner King recorded his reaction:

> [He] thought it was a great blank not having the dates at the top of the pages in the new version. Where they are only guessed at in the very ancient records he would have them put with a question mark, but when we get to the time of Solomon and so on the dates are as accurately known, he says, as that of the battle of Hastings.

This wasn't quite the end for Ussher's chronology; it lingered on in other editions of the Bible for another two decades. Cambridge University Press removed it from their bibles in 1900, and Oxford University Press followed suit in 1910.

★ ★ ★

Kelvin delivering his last lecture at Glasgow University in 1899.

Despite Kelvin's efforts and his limiting timescale of 100 million years, most biologists continued to believe that species had evolved by natural selection. Darwin's creed lived on, and as ever, this remained the true bugbear for Kelvin and his supporters, many of whom argued that Kelvin had actually overestimated the age of the earth. The Scottish mathematician Peter Guthrie Tait, for instance, reworked Kelvin's calculations for a cooling earth and produced a figure of 15 million years. Other scientists reached the same conclusion, and before long there was a clamour for Kelvin himself to re-examine the question.

In June 1897, at a meeting of London's Victoria Institute (a society that strove to reconcile the discoveries of science with the revelations of religion), Kelvin, now in his seventies and sporting a bushy white beard, showed he had lost none of his former enthusiasm for harrying the biologists. In a speech ringing with conviction he declared that the world was much younger than his earlier results had indicated. New values for the specific heats and melting points of rocks showed that the earth had solidified between 20 and 30 million years ago. At the same time, the most recent estimates of the sun's energy indicated it had only been shining for around 20 million years. Gone were the vague estimates of former papers; in their place he spoke of 'strict limitations', of 'certain truth', and 'no possible alternative'. He was pushing Darwin's supporters, and the geologists, into a corner.

It was a step too far. This time they refused to bow to the mathematical precision of physics. Almost to a man, they rebelled. 'There must be some flaw in the physical argument,' cried one. 'I would respectfully submit, there must be some serious error in the data,' argued another. Such a short timespan just didn't fit with the geological or biological evidence; the physicists had to be wrong.

One of the reasons for this sudden flare of dissent was that geologists had begun to investigate the age of the earth using their own methods. During the last quarter of the nineteenth century a number of them had estimated the earth's age by repeating Phillips' method of measuring the rate at which sediment settled on the ocean floor. Of the fifteen or so geologists who tried this technique, at least ten put the age of the world at, or close to, 100 million years.

Furthermore another, completely independent, method produced the same result. In the 1880s, John Joly, the inventive professor of geology at Dublin's Trinity College, revived the idea of measuring the age of the oceans using salt.

His method was similar to the one Edmond Halley had proposed nearly two hundred years earlier, only instead of using common salt (sodium chloride) as a guide, Joly chose to measure just the sodium content, which he considered a more accurate indicator of the ocean's age. Apart from this, the principle was the same. Like Halley, Joly believed that the earth was initially too hot for oceans to form, but as the globe cooled, water vapour in the atmosphere condensed and fell as rain, covering the planet with water. At first this primitive ocean consisted of pure water, but gradually, as rivers dissolved minerals out of the rocks and washed them into the sea, the ocean had become progressively saltier.

Joly, however, took a subtly different approach to Halley; instead of trying to measure directly how much the salinity of the oceans had varied over time – an impossible task, even with Victorian technology – he decided to find the total amount of sodium in the ocean, and divide this by the amount of sodium that arrived each year from the rivers of the world. Assuming that the rivers had always carried the same amount of sodium each year, this would give him the ocean's age.

From published figures for the volume of the oceans, and knowing the fraction of sodium in sea water, Joly calculated that the total amount of sodium in the sea was approximately 14 thousand million million tons. Again from published data, he found that the combined outflow of all the rivers of the world poured nearly 160 million tons of sodium into the sea each year. From this he worked out that the oceans had taken just over 90 million years to reach their present salinity.

Backed by this and other evidence, many geologists thought that Kelvin had been right the first time and that his new date of 20 million years had to be wrong. Their reaction was one of defiance. Having retreated in the face of the physicists once, they were in no mood to do so again without at least questioning the evidence. Some astute geologists began to suspect that the fundamental principles on which the physicists had built their arguments were not as solid as the physicists believed.

The most prescient observation came from Thomas Chrowder Chamberlin, the professor of geology at Chicago

Doomsday Postponed

University. In the summer of 1899, in a direct reply to Kelvin's assertion that the sun had only been shining for 20 million years, Chamberlin challenged the basic assumptions of physics:

> Is present knowledge relative to the behaviour of matter under such extraordinary conditions as obtain in the interior of the sun sufficiently exhaustive to warrant the assertion that no unrecognised sources of heat reside there? What the internal constitution of the atoms may be is yet open to question. It is not improbable that they are complex organizations and seats of enormous energies.

By raising the possibility that atoms might contain massive amounts of energy, Chamberlin anticipated the dawn of the atomic age. Within five years, a growing understanding of the atomic nucleus would turn traditional physics on its head, scatter Kelvin's pronouncements to the winds, and loose the shackles of his restrictive timescale.

The trigger for this sea-change was the realisation that the sun was powered, not by gravity as Kelvin believed, but by a previously unheard-of source of energy – atomic power. The discovery that led to this conclusion had been made in Paris three years earlier. In the spring of 1896 Henri Becquerel, a physicist at the Muséum National d'Histoire Naturelle, spread a thin layer of white crystals of a uranium salt on top of a black-wrapped photographic plate. Between the crystals and the plate lay a thin copper cross. When he developed the plate a few days later, a fuzzy image of the cross appeared; the crystals were emitting a 'type of invisible phosphorescence'. Two years later, Marie and Pierre Curie gave it the name we know today – radioactivity.

At first, Becquerel's discovery attracted little attention. Uranium was hard to obtain, and the feeble rays it produced gave poor, fuzzy images. Most scientists preferred to study the recently discovered X-rays, which produced clearer, sharper pictures. Then, in March 1902, Marie Curie isolated radium, a substance a million times more radioactive than uranium, and everyone sat up and took notice.

By the autumn of 1903, radium was making headlines all over the world. Billed as the most expensive metal on earth, the public lapped up newspaper stories of its fabulous powers. It could kill bacteria, cure blindness, reveal the sex of an unborn child, and turn the skin of Negroes white. A single gram would lift five hundred tons a mile in height, and an ounce could drive a car around the world. Adverts promoted its use in 'radio-active drinking water' for the treatment of gout, rheumatism, arthritis, diabetes and a whole host of other illnesses. By the end of the year, scientists would also claim it as a new source of energy for the sun.

The trigger for this outburst of 'radium fever' was not so much the initial discovery but an experiment Marie's husband Pierre had performed in March 1903. Together with his young assistant, Albert Laborde, he had measured the amount of heat given off by a small flask of radium, and found that a single gram could heat its own weight of water from freezing to boiling in an hour. Even more astonishing, it continued to release heat, day in day out, for months on end without changing its molecular structure. This, in itself, was unheard-of, but there was more. When Pierre cooled the radium down to −180C and heated it to +450C its activity remained constant. Whatever was generating the energy, it was no ordinary chemical reaction. There were only two possible explanations: either the radium was receiving energy from an unknown external source and somehow converting it into heat; or the energy came from within the atoms of radium itself. Exactly where the energy came from became the most important question in physics of the day, but Pierre was unable to answer it.

The mystery was solved by Ernest Rutherford, a young New Zealand physicist working in Canada. Rutherford was just twenty-seven when he took up the professorship of physics at Montreal's McGill University in 1898, but within a short time his irrepressible energy and unbounded talent opened a window onto a whole new area of science – the nature of the atomic nucleus.

However, he had made his first discoveries – the first steps towards understanding why radium continuously poured out heat – a few years earlier in England. At the Cavendish Laboratory in Cambridge he had discovered that radioactive materials emitted at least two types of radiation: low-energy alpha rays, that could be stopped by a few sheets of paper, and the more penetrating beta rays. In Canada he continued his researches, and by 1902 had found the explanation for Pierre Curie's mystery.

Studying a compound of the radioactive element thorium, Rutherford found that it gave off an invisible radioactive 'emanation' – a mysterious odourless gas. Keen to discover what it could be, he asked Frederick Soddy, a young English chemist working at McGill, to carry out an analysis. Soddy discovered that the gas was a previously unidentified inert gas, like the recently discovered argon, but he also found something much more surprising. The gas wasn't being produced by a chemical reaction, but was emanating from the thorium itself. This was almost too incredible to be believed. One of the long-standing laws of science said that elements, such as thorium, could not be split into smaller parts, yet that was exactly what was happening. Soddy had witnessed the alchemists' dream – the transmutation of one element into another.

He later recalled the moment of realisation. Looking up

from the experiment, he called across the laboratory: 'Ruther-ford, this is transmutation: the thorium is disintegrating . . .'

'For Mike's sake, Soddy,' Rutherford shouted back, 'don't call it *transmutation*. They'll have our heads off as alchemists.' All the same, he waltzed around the room in elation, thundering out 'Onward Christian so-ho-hojers' to all who chose to hear.

The discovery was a huge clue as to what was causing Pierre Curie's sample of radium to give off so much heat. Within a short time, Rutherford and Soddy showed that the heavy atoms of thorium, radium and other radioactive elements spontaneously broke down into lighter atoms of other elements. As they did so, they spat out the tiny particles – alpha and beta 'rays' – that Rutherford had noticed earlier. Of these, the larger alpha particles were the main source of heat. Ejected at tremendous speeds – Rutherford measured them at 25,000 km per second – they soon collided with nearby atoms or molecules, transferring their kinetic energy to the surrounding material and raising its temperature. Because radium disintegrated faster than most other radioactive substances, it released more of these particles per second, and therefore produced more heat.

Armed with this information, Rutherford prepared to stick his neck out and answer the question Pierre Curie had left open: where did the energy of radium come from? He was confident that no outside source was involved. Instead, atoms of all elements possessed 'an enormous store of latent energy', and it was this 'internal energy' that enabled radium to produce the continuous supply of heat that Pierre Curie had measured.

The implications were far-reaching. If Rutherford was right, then radioactivity could power the sun – and power it for far longer than the lifetime Kelvin had predicted. In the spring of 1903, he and Soddy wrote a ground-breaking paper in which they announced, with typical bullishness, that 'the maintenance of solar energy . . . no longer presents any fundamental difficulty'.

Despite his faith in this assertion, when Rutherford arrived in England in the summer of 1903 to present his ideas to the British Association he felt uncharacteristically nervous. He knew that Lord Kelvin opposed his views and was likely to put up a formidable fight at the meeting.

Kelvin's opposition was understandable. If radioactivity fuelled the sun, then the earth could be much older than 20 million years. Conceding defeat now would open the way to the acceptance of natural selection. 'I cannot think how Rutherford and others can complacently attribute the emission of heat by radium to energy stored in the substance,' Kelvin told a colleague, 'the hypothesis of *evolution* of an atom seems to me utterly wild and improbable.' The idea of one element changing into another was every bit as horrifying to the classical physicist as the transmutation of species, but that a substance could release more heat than it received appeared to fly in the face of common sense. Kelvin was certain that the heat given off by radium was coming from an outside source.

Meanwhile Rutherford, fearing a confrontation with Britain's leading physicist, had written to a number of friends, asking them to attend the debate and lend their support. What he didn't know, however, was that Kelvin was struggling to come up with a plausible explanation of his own. 'We are all at our "wits' end" in respect to the emission of heat by radium,' Kelvin moaned to a friend, 'and are forced into thinking of, if not actually harbouring, very wild conjectures . . . waves of ether may conceivably supply the energy radiated out by radium; but I *cannot* at present see how they do it; or whence *they* get the energy they conjecturally supply.'

In the event, Rutherford didn't need his planted supporters. In the months leading up to the meeting, his 'disintegration' theory won the support of a large number of British physicists. When the day of the anticipated confrontation arrived, Kelvin pleaded ill health, and stayed away.

It was the turning point. A week after the meeting George Darwin, now professor of astronomy and experimental philosophy at Cambridge, wrote to *Nature* retracting his support for Kelvin's timescale. If the sun could be fuelled by energy from within the atom itself, he explained, then Kelvin's gravitational theory was redundant, and so too his estimate of the sun's age. Darwin calculated that the amount of radioactive 'energy available is so great as to render it impossible to say how long the sun's heat has already existed, or how long it will last in the future.' His desertion was another nail in the coffin for Kelvin's timescale; but worse was to come.

Having demolished Kelvin's argument for the lifetime of the sun, Rutherford now went on to do the same for the earth. Early in 1904, two German physicists, Julius Elster and Hans Geitel, announced that radioactive matter was present in low concentrations virtually everywhere they looked: in well water, hot springs, even ordinary garden soil. When Rutherford read their report, he immediately grasped its significance. The earth's temperature gradient (which Kelvin had used to calculate the planet's age) might not be the result of the earth cooling, as Kelvin maintained, but instead might be produced by radioactive elements generating heat deep below the surface. Intrigued, Rutherford calculated what proportion of radioactive material in the earth would be needed to produce an identical temperature gradient. It came to one part of radium for every 20 million million parts of other material – approximately the amount discovered by Elster and Geitel.

In May 1904 Rutherford returned to London, where the Royal Institution invited him to deliver one of their popular Friday night lectures. He chose the occasion to deliver the knock-out blow to Kelvin's timescale. By now Rutherford's discoveries had made him famous, and his presence drew the crowds. Over eight hundred people pushed their way through the Institution's doors in Albemarle Street, and up the stairs to the packed lecture theatre. It was the largest audience for five years. Among them sat Kelvin. Rutherford recalled what happened next:

> I came into the room which was half dark, and presently spotted Lord Kelvin in the audience and realized I was in for trouble at the last part of my speech dealing with the age of the Earth where my views conflicted with his. To my relief Kelvin fell fast asleep, but as I came to the important point, I saw the old boy sit up, open an eye and cock a baleful glance at me! Then a sudden inspiration came and I said that Lord Kelvin had limited the age of the Earth *provided no new source* [of heat] *was discovered.* 'That prophetic utterance refers to what we are considering tonight, radium!' Behold! the old boy beamed upon me!

Kelvin didn't smile for long. As Rutherford brought his lecture to a close, he chose his last sentence to breathe new life into Darwin's theory – the theory that Kelvin had spent the last forty-five years trying to bury. 'The discovery of the radio-active elements . . .' Rutherford concluded, 'thus increases the possible limit of the duration of life on this planet, and allows the time claimed by the geologist and biologist for the process of evolution.'

For once good news rolled off the presses. 'DOOMSDAY POSTPONED' screamed the headline in at least one newspaper the next day, as it pointed out that the new source of heat would mean a longer-lived earth.

The following weekend, William Strutt, the Cambridge professor of physics, invited Rutherford and Kelvin to a house party at his country home in Essex. Despite his best efforts, Rutherford was unable to persuade the old man to change his view that radium was simply re-emitting energy absorbed from some outside source. 'Lord Kelvin has talked radium most of the day,' he reported that evening in a letter to his wife, 'and I admire his confidence in talking about a subject of which he has taken the trouble to learn so little. I showed him and the ladies some experiments this evening, and he was tremendously delighted and has gone to bed happy with a few small phosphorescent things I gave him. He won't listen to my views on radium, but Strutt gives him a year to change his mind. In fact they placed a bet to that effect last evening.'

According to one report, Kelvin conceded the five-shilling bet, and paid up later that year. In public, however, he never abandoned his conviction that radium was simply re-emitting energy absorbed from some outside source; nor did he abandon his belief in a short-lived world. In a series of letters to *The Times* written in 1906, the year before he died, he categorically denied that radium supplied the heat of the sun or the earth, and continued to maintain that only gravitation could provide such vast reserves of energy.

It was his final pronouncement on the subject.

11. Rock of Ages

The year 1904 marked a turning point in the search for the beginning of time. During the summer Rutherford found what philosophers had dreamed of for centuries – a natural chronometer that could accurately date the earth. Radioactivity not only heated the planet; it could determine the age of the rocks buried in it. Exactly when he made the discovery isn't known, but his timing could not have been better. The organisers of a prestigious international congress had invited him to the World's Fair in St Louis that September to address a gathering of the world's leading scientists, and he wanted something special to announce to match the grandeur of the occasion. Dating rocks with radioactivity would be it.

In its early, crude, version, Rutherford's method was based on measurements of a newly discovered gas – helium. In the early 1890s William Hillebrand, a chemist working for the American Geological Survey, had noticed that when he put minerals containing uranium, such as clevite, into solution, they released a large quantity of gas. In a hurry, and too busy to carry out a full analysis, he had assumed the gas was nitrogen and described it as such in his report, and that would have been the end of the story had his paper not been spotted by Sir Henry Miers, Keeper of Minerals at the British Museum. Miers, suspecting that the gas might not be nitrogen at all, but the recently discovered inert gas, argon, wrote to

Sir William Ramsay, the co-discoverer of argon, informing him of his suspicion. Within weeks Ramsay had bought a gram of clevite from a dealer, extracted the gas and discovered, not argon, but another inert gas that had never been seen on the earth before – helium.

How the helium came to be in the rock turned out to be an even more fascinating story, and it was this that provided the key to Rutherford's dating technique. As early as 1902 Rutherford suspected that the alpha particles he had observed flying out of thorium and other radioactive elements might actually be positively charged atoms of helium. But he couldn't be completely sure. The fact that helium had been found in uranium-bearing rocks made it very likely, but this wasn't enough for proof. His colleague, Soddy, decided to investigate the matter. In February 1903, he left Montreal to work alongside Ramsay in London. Together they planned an experiment to see if radium gave off helium.

That summer, they placed twenty milligrams of radium bromide in a small flask, added water, and drew the resulting gaseous mixture through a delicate rig of glass tubing. In the darkened room they watched the progress of the gas, glowing with radioactivity, as it passed slowly through the thin capillary tubes, rushed along the wider connecting pipes and finally tumbled into a U-tube where it was cooled by liquid air. As it condensed, the glow became brighter. But Soddy and Ramsay were not interested in this shimmering liquid (which later turned out to be radon). Another gas, that hadn't condensed, remained in the tube. This, they believed, was helium, but the only way to find out was to examine its spectrum.

It had recently been found that when an element, such as helium, was heated, it emitted light at characteristic wavelengths. Viewed through a spectroscope – a prism, or fine grating that spreads white light out into the rainbow of colours from which it is made – these wavelengths appear as distinct coloured lines that uniquely identify that element, just as surely as a fingerprint identifies a criminal.

Soddy fed a tiny trace of the gas into a small vacuum tube which could be heated by an electric discharge. 'I gave Sir William the all clear, and he switched on the coil and glanced with his pocket spectroscope at the spectrum tube, exclaiming at once in a tense voice, "That's D_3" [the tell-tale line of helium]. As I worked, the room began to fill silently, as if by some telepathic process the news of the success of the experiment began to spread through the laboratory.' Staff and students crept into the room, until there was barely space to stand. 'The spectroscopes passed from hand to hand . . . I was entirely absorbed in the complicated gas manipulation . . . completely forgotten, an alien among a band of brothers: and I suppose I was about the last in that crowded room to see that single bright yellow line.'

When the news reached Canada, it persuaded Rutherford that alpha particles were indeed positively charged atoms of helium. This meant that any radioactive element that gave off alpha particles, for instance uranium, also produced helium. Combined with the discovery that uranium-bearing rocks contained measurable quantities of helium, it was enough to convince Rutherford that his method of dating rocks would work. On 13 September 1904 he packed his bags, and caught the train to St Louis, to reveal his idea to the world.

The International Congress was a side-show to the most exciting event in America that year: the St Louis World's Fair – the largest international exposition the world had ever seen. 'Meet Me in St Louis' went the popular song of the time, and they did in their legions. Over nineteen million visitors flocked into the city to see the razzle-dazzle of the show. Strolling through a vast park dotted with wedding-cake buildings, they gazed at the latest wonders of the world: the wireless telegraphy tower, sending messages through the ether; the Palace of Electricity, housing such modern marvels as a telephone exchange; and the Palace of Transportation, chock-full of electric trains, streetcars and over 140 automobiles. Outside, the skies were filled with primitive airships, as inventors competed to surpass the Wright brothers' achievement of the previous year, and fly a figure-of-eight course. The crowds listened to the marches of John Philip Sousa, or danced the hootchy-kootchy; they bought peanuts and chewing-gum from the new vending machines and for the first time ate ice-cream from a cone. At night the buildings were lit up in electric lights. Here were the inventions that would shape the twentieth century, a technological century, America's century. Rutherford slipped away from the congress to drink it all in. 'Very fine,' he remarked on the illuminations. The scientific centre of gravity was leaving Europe and heading west.

It was Soddy who saw the darker side of it all. Years later he recalled seeing at the fair a 'gigantic human figure in cast-iron many times life-size representing the god Vulcan. Along-side of it . . . was a diminutive exquisitely carved figure in white marble . . . of the Christ'. The juxtaposition, with 'its obvious moral of the dominion of the spiritual by the material', appeared to anticipate 'the whole world about to be subjugated by atomic energy'.

The congress was packed. They came from far and wide – Japan, Russia, Europe, Mexico. Among them, the most eminent scientists of the day. With so many people to speak to, and talks to attend, Rutherford had a hard time fitting it all in. On the Thursday afternoon he finally presented his paper.

His idea was that the amount of helium in a rock could be used to measure its age. He knew that helium was always found in rocks containing radium. He also knew that radium emitted helium – probably in the form of alpha particles – at a steady rate. From this it followed that the gas would steadily accumulate in the rock, and older rocks would contain a higher proportion of helium than younger ones. Assuming you knew the rate at which helium was produced, you could find the age of the rock.

This assumed, of course, that there wasn't any helium in the rock in the first place. But it seemed unlikely that there would be. While the rock was still in its hot molten state, any helium present would have bubbled out, escaping into the atmosphere. Only after the rock solidified would the helium have become trapped.

The difficult part had been determining the rate at which helium was produced. From a delicate experiment based on measuring the charge of alpha particles emitted from radium, Rutherford had calculated how many helium atoms were produced per second, and from this had deduced that 1 gram of radium would release 200 cubic millimetres of helium per year.

The audience of eminent scientists listened intently as Rutherford spelt out his conclusion. While the necessary measurements had 'not yet been determined with sufficient accuracy to make . . . more than a rough estimate of the age of any particular mineral', he had little doubt that, when they were known, the method could 'be applied with considerable confidence'. Meanwhile, he had already made the first 'rough estimate'. Measurements of the amount of helium in a piece of Norwegian rock – a black vitreous mineral called fergusonite – put its age at almost 40 million years.

The great beauty of this technique, as Rutherford realised, was the reliability of radioactivity as a timekeeper. As Pierre Curie had shown with his experiment on radium, radioactivity was unaffected by changes in temperature and pressure. However hot or compressed the rocks became, helium would continue to be produced at the same rate; the radioactive clock would keep on ticking. In this respect, radiometric dating, as this method came to be called, represented a huge advance over earlier geological methods based on changes in fossil species, or the thickness of ocean sediments.

The following year, using improved data, Rutherford recalculated the age of the same sample of rock and declared it to be 500 million years old. The world was getting older again. But not for long.

By the middle of 1905 it became clear that the helium method suffered from a fundamental flaw. Over time, the gas slowly seeped away from where it was made and dispersed into the surrounding soil. As some minerals were more porous than others, there was no way of knowing how much helium had escaped. This meant that the amount of helium measured would be less than the total amount produced. The technique could give the minimum age of a rock, but it couldn't give its true age. To find that, Rutherford realised he needed another method – and he needed help.

Like Kelvin, Rutherford tended to be scathing about other branches of science. Physics, of course, was the true science, the king to mathematics' queen; but chemistry was 'stinks'; while the rest – such as geology and biology – he dismissed as mere 'stamp-collecting'. Without the stinkers and stamp-collectors, however, his attempts at radioactive dating would have got nowhere. Rutherford always needed good chemists to help him in his research. In the past Soddy had filled the gap, but with his departure for London he needed a replacement, a careful, methodical experimenter whose results he could trust. During a visit to Yale University in New Haven, Connecticut, in the spring of the previous year, he had bumped into just the man.

Bertram Borden Boltwood was thirty-three when he met

Rock of Ages

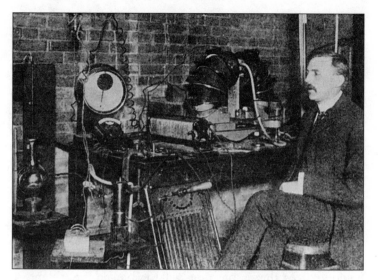

Ernest Rutherford at McGill University in 1905.

Rutherford at Yale. Tall, with a mischievous sense of humour and a love of good food and fine wines, he worked in the laboratory of the New Haven mining business he ran with his partner, Joseph Pratt: 'Pratt and Boltwood, consulting mining engineers and chemists'. While Pratt tramped the country collecting ore samples, Boltwood stayed in New Haven, analysing the rocks his partner sent back. Many of these happened to contain uranium and thorium, and Boltwood naturally became interested in their radioactive properties. After his meeting with Rutherford, the two struck up a correspondence, answering each other's scientific queries. Through these letters, Boltwood led Rutherford to a better way of dating rocks.

From his experience of analysing uranium minerals, Boltwood noticed that they always contained lead. Could this lead have come from uranium? he asked. It seemed likely. By now it was known that uranium passed through several stages as it decayed. At one point it transformed itself into radium, but no one knew what element was at the end of this cascading chain; Boltwood suspected it was lead.

He also realised how he could check this hypothesis. If lead was the final product of uranium it would accumulate in the rocks: older minerals would contain a higher proportion than younger ones. He looked back through Hillebrand's analyses of uranium-containing rocks, and found that this was indeed the case: the older the rock (determined by its position in the strata), the more lead it contained.

Carried away in his enthusiasm, he wrote to Rutherford: 'If lead can be shown to be a disintegration product of uranium, will it not necessarily follow that all the lead existing on the globe originated in this way? I think the deductions which can be made from this assumption will make even the metaphysicians dizzy.'

Having determined that lead was the final disintegration product of uranium, it didn't take Rutherford and Boltwood long to realise that it would make a better indicator of a rock's age than helium. Unlike helium, lead didn't diffuse out of the rock, so the amount they measured was much more likely to be the total amount produced. If he and Boltwood could determine the rate at which lead was made, then, by measuring the relative amounts of uranium and lead in a rock, they could calculate its age.

This wasn't easy. Uranium decayed so slowly that the amount of lead produced was impossible to measure directly. Nevertheless, with Boltwood's help, Rutherford managed to determine it indirectly. It was fantastically slow; in the course of a whole year less than one uranium atom in 2 billion decayed to lead. The amount of lead produced was minuscule.

Yet the proportion of lead found in uranium-bearing minerals was quite significant; it had obviously accumulated over a considerable time. In the autumn of 1905, using the new lead method, Boltwood calculated the ages of twenty-six samples of rock. He found the youngest was 92 million years old, while the oldest, a piece of thorianite from Ceylon, dated back 570 million years. When he checked the dates with his geological colleagues, he was delighted to find that they agreed perfectly with the relative ages of the different deposits. The method worked. He revealed his results in a letter to Rutherford, but strangely didn't publish them, possibly thinking they were still too speculative.

It was just as well. Two years later when he recalculated his figures using a new estimate for the rate of uranium decay, he found the ages had shot up. The youngest rock was now 410 million years old, while the oldest was a staggering 2,200 million years. This time he published his results. It was barely seven years since Rutherford had first challenged Kelvin's brief timescale of the world, yet in that time its age had increased dramatically. For a public used to dealing in everyday numbers of tens and hundreds, this huge jump, from millions to billions, was dazzling.

After this initial flurry of activity, Rutherford and Boltwood drifted away from radioactive dating. Boltwood returned to Yale, where he discovered a new element – ionium; while Rutherford moved back across the Atlantic, to Manchester University, where he built up a formidable team of researchers, turning it into one of the foremost laboratories for nuclear research in the world. Just before the outbreak of the First World War, however, he made a further excursion into the question of the earth's age.

Although the technique he developed is no longer used, it is worth describing what must surely rank as the most charming and unusual method of dating the earth ever devised. Sadly it was also one of the least successful. It was based on the colour of haloes; not the angelic type, but microscopic ones that appear in rocks. When thin sections of mica and certain other minerals are viewed through a microscope, small disc-shaped marks known as pleochroic haloes are sometimes seen. For years geologists had puzzled over what could have caused these strange microscopic spheres, but in 1907, John Joly – the Irish geologist who had estimated the age of the oceans from their salt content – found the answer: they all contain a tiny radioactive crystal – normally zircon – at their centre. As radioactive elements in the zircon decay, they throw out alpha particles in all directions, which penetrate the surrounding rock and change its colour. How far the particles travel depends on the element from which they originate – those produced by the decay of radium, for instance, are thrown further than those from uranium and therefore create a larger sphere. However, the most interesting thing Joly found was that the colour of the haloes varies with the age of the rock. The more alpha particles thrown off, the darker the rock becomes. The mica behaves like a very insensitive photographic film, needing an exposure of thousands of years to record an image. The older rocks therefore have the more noticeable haloes.

Together with Rutherford, Joly attempted to estimate the age of a piece of Devonian granite from the colour of its haloes. Natural haloes are made with minuscule amounts of radiation released over a long period of time, but Rutherford realised he could make artificial ones quickly in the laboratory using stronger sources of radioactivity. Under controlled conditions, he bombarded fragments of the same rock with measured amounts of radiation to create a series of artificial haloes. He then compared their colour with that of the real haloes. Once a match was found, he measured the amount of radioactivity he had used to make the artificial halo, and compared it with the radioactivity of the zircon at the centre of the real halo. From this he was able to estimate the age of the rock. Tests on several haloes gave results that varied between 50 and 470 million years.

Curiously, despite this collaboration, Joly stubbornly refused to accept that dates based on radioactive decay were accurate. He argued that radioactive decay had proceeded faster in the past than at present, and clung tenaciously to his estimates based on the salinity of the oceans. He wasn't alone. It would be some years before many of the old-school geologists could shake off Kelvin's dates.

With the outbreak of the First World War, the team of researchers Rutherford had carefully gathered together at Manchester broke up. Never before had scientists been so caught up in warfare. Colleagues who had collaborated in the laboratory months earlier now found themselves fighting on opposite sides of the battlefield. Those more fortunate were seconded to military research behind the lines. Rutherford was detailed to work for the Admiralty developing ways to detect enemy submarines, while across the water in Ireland, Joly found himself at Easter 1916, armed with a revolver, defending Trinity College from an attack by Sinn Fein.

As the rebellion against British rule turned the centre of Dublin into a battlefield, Joly and a handful of others manned the walls of the university ready to fend off any assault. The situation was serious. On the second day, one of his fellow defenders shot a suspected rebel messenger who had made the mistake of cycling past the buildings. They dragged his body into the college precinct and buried him in the grounds. But Joly felt he was defending more than British rule. At the heart of the university stood Ussher's library, with its valuable collection of ancient manuscripts. As he later recalled,

... had the Sinn Feiners taken possession of the College buildings in force, nothing but shell-fire would have dislodged them. Having regard to the great strength of the place, no other course but one which must probably involve the ruin of the buildings would be justifiable . . . With the Library, enriched at the voluntary expense of soldiers, the most precious heirlooms of ancient Irish civilisation would perish . . .

Although the uprising was crushed, Joly was fighting for a lost cause; the rebellion marked the beginning of the end for British rule in Ireland.

He was losing a battle on another front as well. After the war, radioactive dating techniques evolved steadily, thanks largely to the work of Arthur Holmes, a young physicist working at Imperial College, London. Almost single-handedly, Holmes refined the lead method until, in 1921, it was acknowledged as the preferred method of geological dating. In the 1920s radiometric ages finally began to replace those of Kelvin in the textbooks. In his own book *The Age of the Earth*, published in 1927, Holmes announced that: 'All the evidence is consistently in harmony with the conclusion . . . that the age of the earth is between 1,600 and 3,000 million years.'

But, as Holmes himself realised, the search for the dawn of time was about to enter a new phase. Having ploughed the depths of the earth, scientists now turned their attention to the skies. Space became the new frontier to be explored. It was a vaster, less tangible arena, but it held out the promise of ultimate success. In his book's evocative concluding paragraph, Holmes foresaw the possibilities:

Beyond the earth, receding into an inconceivable remoteness, lies the stellar universe, the stage in which the drama of our earth is set. Here another veil is lifting, for Nature, no longer taciturn when approached in the right spirit, is slowly but surely yielding up her secrets. The nature of the evolution of the stars and the hidden sources of their blazing energy are being revealed; and we are adventuring towards the very threshold of Creation – towards the birth-time of the universe itself.

12. Star Gazing

The great philosophical leap that brought astronomy into the search for the beginning of time had actually occurred over two hundred years earlier. In 1676, the Danish astronomer Ole Rømer announced to the disbelieving members of the *Académie des Sciences* in Paris that light took a finite time to travel through space. Before then most philosophers believed that vision was instantaneous: when they saw an event happen, however distant, they believed they were seeing it at the very moment it occurred. If Rømer's announcement was true, however, it meant that astronomers, instead of seeing the flash of an exploding star at the instant it occurred, were watching it hours, or even years, after the event. When they looked out into space, they were looking back in time.

Rømer's discovery came from a brilliant piece of reasoning. Like other astronomers of the day, he knew that Io, one of Jupiter's moons, circled Jupiter approximately every 42 hours, sweeping around the planet in much the same way that a minute hand sweeps around a clock. He also knew that, unlike a minute hand, Io didn't keep perfect time. Sometimes it disappeared into Jupiter's shadow later than expected; at other times the eclipse occurred early. To find out why this should be, he observed the moon for several years; every time Io disappeared into Jupiter's shadow, he noted the date and time. Eventually he realised that something curious was happening.

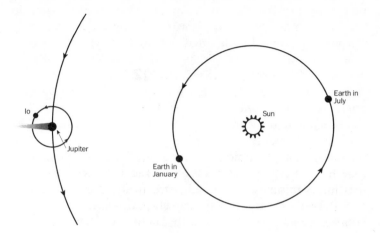

How Rømer determined the speed of light. Light from Jupiter's
moon, Io, arrives earlier when the Earth is nearest Jupiter (January)
than when it is furthest away (July). The speed of light can
therefore be calculated from the time it takes to cross the
diameter of the Earth's orbit.

Io's eclipses were earliest when the Earth was on the side of the
sun closest to Jupiter; as the Earth continued on its orbit and
moved away from Jupiter, they grew progressively late. Rømer
realised that this delay was due to the fact that light took time to
travel through space. When the Earth was at its greatest distance
from Jupiter, the light from Io would have to travel further,
and would therefore take longer to arrive. Going through his
tables of eclipse times he worked out that it took 22 minutes
for the light from Io to cross the diameter of the Earth's orbit.
This was the first estimate of the speed of light.

His figure was soon improved. In 1704, Newton wrote in

his *Opticks* that 'Light is propagating in luminous bodies in time, and spends about seven or eight Minutes of an Hour in passing from the Sun to the Earth.' And by 1729 the English astronomer James Bradley had narrowed down this journey time even further, to 8 minutes and 12 seconds – within 3 per cent of the present-day value.

Rømer's discovery launched a revolution in the way astronomers viewed the universe. Before it, they believed that they were seeing all the stars and planets at the same instant – the present. After it, they realised that the further away the object, the further back in time they could see. When they looked at the Moon, they saw it as it was one-and-a-half seconds earlier, the Sun 8 minutes 12 seconds earlier, and Neptune a little over 4 hours earlier. But these were Earth's intimate neighbours. The stars were much further away, and their light must have taken much longer to reach us. The realisation of this opened a new era for chronology. By measuring the distance to remote stars, astronomers could determine how long ago the light had left those stars, and thereby find a minimum age for the universe.

In 1813 the German-born astronomer William Herschel, working in England, found a way of measuring distances to remote stars. His results amazed his contemporaries. 'I have looked further into space than ever human being did before me,' he told Thomas Campbell, a visitor to his observatory. 'I have observed stars, of which the light, it can be proved, must take two millions of years to reach the earth . . . Nay, more . . . if those distant bodies had ceased to exist two million years ago, we should still see them, as the light would travel after the body was gone.' The revelation astounded Campbell, who later wrote: 'I really and unfeignedly felt at this moment as if I had been conversing with a supernatural intelligence.'

Unfortunately Herschel's estimate was horribly wrong – he had based it on the false premise that stars, like candles, all shone with the same absolute brightness, and that the further away the star, the dimmer it would appear. Because the apparent brightness of an object declines in a strict mathematical ratio to its distance, Herschel thought he could estimate how much further one star was than another, just by measuring how brightly it shone. Sadly he was mistaken. Astronomers soon discovered that there were many different types of star, some millions of times brighter than others. Without being able to tell one sort from another, his method was doomed to failure.

In principle, however, Herschel's idea was a good one. Conventional geometrical methods were only capable of determining the distances to the nearest handful of stars; to find the distance to remote stars, astronomers had to rely on measurements of brightness. If they could find a type of star that always shone with exactly the same absolute brightness, a 'standard candle' as it came to be known, and could find a way of picking it out from among all the other stars in the night sky, they would have a way of measuring deep into space.

The much-needed breakthrough came in 1912 from a quiet, conscientious astronomer named Henrietta Leavitt. Leavitt was one of a small team of women 'computers' employed by Harvard College Observatory in Massachusetts to analyse and catalogue the thousands of photographic plates the observatory's telescopes produced each year. For seven hours a day, six days a week, she, and around a dozen other women, sat in a room and plotted the positions of stars, analysed their spectra, and measured their brightness. It was painstaking, detailed work, but its visual nature suited Leavitt, who was profoundly deaf. Her abilities were soon recognised, and before long she was promoted to head the photometry department.

It was her single-minded dedication to her work, however,

Aeons

Henrietta Leavitt.

that led to the important discovery. Early in 1904 she began examining a shipment of photographic plates sent back from Harvard's southern telescope at Arequipa in Peru. Among them were photographs of the Large and Small Magellanic Clouds, dense patches of stars in the southern sky named after the Portuguese explorer who first brought them to the attention of European scholars. Leavitt began scanning them for variable stars – stars that dim, then brighten again, usually in regular cycles lasting from a few days to several months. To find them she had to compare the negative plates (on which the stars appeared as tiny black dots) for stars that showed signs of change from one night to the next. It was slow work, but by the end of the year she had found 152 variables in the Large Magellanic Cloud and 59 in the Small. The following year brought even greater success with the discovery of a further 843 variables, all in the Small Magellanic Cloud. 'What a variable-star "fiend" Miss Leavitt is,' wrote a Princeton professor, impressed by her prodigious output. 'One can't keep up with the roll of the new discoveries.'

Amongst all these stars, however, there was one particular type that attracted Leavitt's attention: the bright variables known as cepheids.

Cepheids were among the brightest stars in the galaxy. (Although their peak brightness varies, an average cepheid is now known to be about ten thousand times brighter than the Sun.) They had been discovered over a hundred years earlier by John Goodricke, a young English astronomer who noticed that Delta Cephei, the fourth-brightest star in the constellation of Cepheus, brightened and dimmed in a five-day cycle. Like Leavitt, Goodricke also happened to be deaf, and perhaps it was this affinity that led her to take so much interest in the stars he had discovered. If so, it was certainly fortunate, for in 1908 Leavitt made an important observation: the brightness of cepheids varied with the length of their cycle. By studying a number of cepheids, with cycles between 1.25 days and 127 days, she found that the longer the cycle, the brighter they shone.

This was intriguing, but of far more importance was the discovery she made four years later. After a detailed study of 25 cepheids in the Small Magellanic Cloud, she discovered that a cepheid's brightness and its length of cycle were linked by an exact mathematical relationship. At its simplest, this meant that all cepheids with the same length of cycle shone with equal brightness; they were standard candles. If one appeared fainter than another, it was because it lay further away. It was now possible to apply the method Herschel had used. By measuring the brightness of the distant cepheid and comparing it to that of the nearer one, astronomers could determine the relative distance of the two stars.

But Leavitt also realised that astronomers weren't restricted to comparing distances between the few cepheids that shared the same length of cycle. Because cycle and brightness were linked mathematically, if the absolute brightness of just one cepheid was known, then the brightness of any other could be determined. In short, all cepheids could be used as standard candles. She had discovered the 'cosmic ruler' astronomers had dreamed of: a way of probing out far beyond the confines of the local stellar neighbourhood.

Before Leavitt's discovery, astronomers had a largely two-dimensional view of the heavens – apart from a handful of the nearest objects, they had no idea of how far away the stars were. Over the next twenty years, using cepheids as a measuring stick, the universe was revealed in its full three-dimensional glory.

The biggest discovery based on cepheid measurements, and the most radical change in the view of the universe, was made by the most famous of twentieth-century astronomers: Edwin Hubble. From the age of eight, Hubble had his heart set on becoming an astronomer. 'Astronomy is something like the ministry,' he once said. 'No one should go into it without a "call". I got that unmistakable call, and I knew that even if I were second rate or third rate, it was astronomy that mattered.'

Although born in Missouri, Hubble appeared for all the world like an upper-class English gent. After university in Chicago, he won a Rhodes scholarship to study at Oxford in England. While there, he shed all trace of his Missouri roots and reinvented himself as an Englishman. He took to wearing tweed jackets, smoked a briar pipe, and rapidly acquired the English accent that stayed with him for the rest of his life. For years to come his exclamations of 'Bah Jove' and 'What ho' astonished everyone he met.

In 1919, after a spell in the army, Hubble arrived in Pasadena in California to take up a post at Mount Wilson Observatory. Situated ten miles north-east of Los Angeles, Pasadena had recently shaken off its agricultural roots, and, fuelled by tourism, had blossomed into a small but bustling community. The observatory offices were situated in Santa Barbara Street, near the centre of town, but it was at Mount Wilson, a winding twenty-mile drive to the north, that Hubble made his great discoveries.

Nearly 6,000 feet high, Mount Wilson commanded a fine view of the Santa Barbara valley stretching into the distance far below; but it was the view heavenwards that everyone came for. The air above the telescopes blew in from the Pacific across four thousand miles of smooth, empty ocean. Free of dust and virtually free of turbulence, it offered the best 'seeing' in North America. For forty years, from 1908 until after the Second World War, Mount Wilson reigned supreme as *the* centre for observational astronomy. There was no better place for viewing the stars.

Edwin Hubble (left) at the 100-inch Hooker telescope on Mount Wilson.

Aeons

In pride of place on the mountain, enclosed in a gleaming white dome, stood the telescope Hubble couldn't wait to lay his hands on – the 100-inch Hooker, the most powerful in the world. Mounted at its base was the huge mirror that gave it its name: a five-ton disc of glass, one foot thick and 100 inches in diameter, that had taken the optical team in Pasadena almost five years to grind to a perfect parabola. Projecting upwards from this was the telescope itself, a 40-foot cylindrical steel cage holding the subsidiary mirrors that reflected light to the eyepiece and camera. Although weighing over 100 tons, it was supported on a bed of mercury and could be moved with the touch of a hand. In operation, however, it used a 3½-ton drive mechanism to ensure it accurately tracked the stars.

When Hubble began working with the 100-inch telescope in December 1919, it was with the intention of studying the structure and shape of hazy Catherine-wheel-like patches of light known as spiral nebulae. His main aim was to learn something of how they evolved, but he also had half an eye on settling a long-running debate. Although astronomers had known about spiral nebulae for over a century, no one knew what they were. Some astronomers, adopting a theory proposed by the German philosopher Immanuel Kant, thought they were giant discs of stars like our own galaxy, the Milky Way, only so far away that their individual stars were no longer discernible. They called them 'island universes' because they resembled remote starry islands floating, it seemed, in the black ocean of space. Others had a more prosaic explanation. They thought nebulae were swirling clouds of gas lying close to, if not within, the Milky Way itself. These two opposing views offered two contrasting pictures of the universe. In the first our galaxy was just one of millions, if not billions of galaxies scattered through space, and space itself was unimaginably large. In the second, the Milky Way comprised the whole visible universe, beyond which lay only dark, empty space.

When Hubble arrived at Mount Wilson most astronomers working on the mountain favoured this second view. They were influenced in part by the opinion of Harlow Shapley, the observatory's most respected astronomer. Shapley had made his name by calibrating Leavitt's distance scale. (He had done this by using conventional geometrical techniques to measure the distances to 11 of the nearest cepheids.) He had then used cepheids as measuring sticks to measure the diameter of the Galaxy. At the time, his result of 300,000 light-years was considered so large that many astronomers, including Shapley himself, considered it must constitute the entire visible universe.

If Shapley was right, however, then chronology based on measuring distances to stars had reached a dead end. If there were no visible objects beyond the Milky Way, then all that could be said was that the oldest stars were at least 300,000 years old, hardly old at all compared to the mind-boggling ages proposed by the geologists.

Hubble, however, kept an open mind. He held no great affection for his respected colleague, and the feeling was mutual. 'Hubble and I did not visit very much,' wrote Shapley, some years later. '. . . Hubble just didn't like people. He didn't associate with them, didn't care to work with them.' It has been said that the two men fell out when Shapley undertook a project that Hubble had hoped to work on. Whatever the reason, relations between the two astronomers remained cool even after Shapley left Pasadena in 1921 to take up the directorship of Harvard Observatory.

Despite Shapley's departure, almost everyone at Mount Wilson continued to believe his assertion that nebulae were nearby objects within the Milky Way, and this was still the dominant view among astronomers when, one day in the fall of 1923, Hubble threw his suitcase into the back of the observatory truck and headed up the mountain to begin another run of observations. The conditions on the night of 5 October were good: the sky was clear and the new moon only three days away. It was dark enough for a nebula shoot. Hubble trained the 100-inch telescope onto the Andromeda nebula, a bright, Zeppelin-shaped spiral in the Andromeda constellation, and ran off a 45-minute exposure. The previous evening he had spotted what looked like a nova – a brief flare-up of a star – in Andromeda, and wanted another plate to make sure. Early the following morning he developed the plate in the mountain darkroom as usual, and examined it closely. Now, instead of just one star flaring up, there appeared to be three. Three novae in one night. This was good hunting.

Back at the Santa Barbara Street offices he rummaged through the library and unearthed plates of Andromeda going back over fourteen years. While two of the stars were definitely novae, the third was altogether more interesting; it appeared to fade and dim in a regular cycle. When Hubble plotted its varying brightness on a graph he instantly recognised the characteristic curve of a cepheid.

The key to the universe had fallen into his hands. Right there, in the stack of plates in front of him, was everything he required to measure the distance to Andromeda and settle the mystery of the spiral nebulae. Shapley had already worked out the tricky details of the relationship between the distance and brightness of cepheids; all Hubble had to do was slot his data into Shapley's formula. When he did this, the result left no room for doubt. Andromeda was a million light-years away – far beyond the furthest reaches of the Galaxy. At that distance, it had to be enormous, a whole galaxy in its own right. Kant had been right. The universe was vast, far, far bigger than Shapley had imagined, and our galaxy was just one among billions.

The Andromeda nebula.

Hubble's excitement must have been intense, but when he broke the news to Shapley, he did so with deliberate understatement. 'You will be interested to hear,' he wrote in a letter dated February 1924, 'that I have found a Cepheid variable in the Andromeda Nebula.' Shapley was not 'interested', he was devastated. Cecilia Payne, a graduate student who happened to be in his office at the time the letter arrived, recalled that after reading it, Shapley passed it to her, saying: 'Here is the letter that has destroyed my universe.'

There is an ironic twist to this story, which suggests that, if he hadn't been so stuck in his opinions, Shapley could have made the discovery for himself. Some three years earlier, while still at Mount Wilson, he was said to have handed some plates of the Andromeda nebula to Milton Humason, one of the night assistants, to examine. Humason, a meticulous observer, noticed what appeared to be variable stars on the plates. He marked their position with a pen, and pointed them out to Shapley, with the suggestion that they were cepheids. Shapley dismissed the proposal out of hand. They couldn't be cepheids, he told Humason, because Andromeda was a nearby object and any cepheids in it would appear much brighter than the stars Humason had pointed out. He took the plates out of Humason's hands, pulled out his handkerchief and wiped the marks clean.

By the end of 1924 Hubble had found 11 new cepheids in Andromeda, and a further 22 in another nebula known as M33. When he calculated their periods, they all gave the same result – both nebulae were almost a million light-years away. On New Year's Day 1925, he revealed his results to the American Association for the Advancement of Science at their winter meeting in Washington, DC. For the first time, the world glimpsed the sheer scale of the universe. Our Milky Way was just one among millions, probably billions, of galaxies scattered through the vast emptiness of space. Kant's vision of scattered 'island universes' floating in a sea of darkness was a reality. If light from the nearby galaxy of Andromeda had taken a million years to reach us, how long must it have taken to reach us from the furthest galaxies? The universe was not only enormous; it was also enormously old.

13. Expanding Horizons

Despite Hubble's success, there was still one mystery about nebulae that remained. And it was quite a mystery. Thirteen years earlier, Vesto Slipher, a quiet, cautious astronomer based at the Lowell Observatory in Flagstaff, Arizona, had set the astronomical community abuzz with an astonishing discovery. Nebulae were flying through space at lightning speeds – some as fast as 1,100 kilometres per second.

Slipher's specialty was spectroscopy, the same technique with which Ramsay and Soddy had identified the helium given off by radium a few years earlier. As well as being a brilliant way to identify elements in the laboratory, spectroscopy could also be used to discover how fast an object was travelling away from or towards the observer.

When an object travels away from an observer, its light appears redder than it would if the object were stationary. This is because the wavelength of the light has lengthened; it has been shifted towards the red (long wavelength) end of the visible spectrum. Conversely, when an object travels towards an observer, its light appears bluer than normal. These differences are subtle, too minute to be noticed with the naked eye, but a similar effect occurs with sound, and can be heard when a police-car with a wailing siren zooms past. The high tone, as the car comes towards you, drops to a lower pitch once it has passed and is speeding away. If you know the true pitch of the siren, then, by measuring how much lower it is as it speeds away, it's possible to determine how fast the car is travelling.

The same principle can be used to measure the speed of a light-emitting object such as a nebula. And this is where spectroscopy comes in: it measures how much the wavelength of light has shifted. Elements in the cool gas surrounding a distant galaxy absorb light at specific wavelengths, so when light from the galaxy is passed through the prism of a spectroscope, the resulting spectrum consists of a series of parallel black lines corresponding to these absorption wavelengths. The spectrum looks a little like a modern bar-code. For a stationary galaxy, any particular absorption line will always appear at a certain fixed wavelength. If the galaxy is moving away from the observer, however, the wavelength of this and all the other absorption lines will be shifted towards the red end of the spectrum. Conversely, if the galaxy is moving towards the observer, they will be shifted towards the blue end. By measuring the amount of this displacement it is possible to determine the change in wavelength of the light and consequently the speed at which the galaxy is moving.

Back in 1910 when Slipher started work, capturing the spectrum of a spiral nebula posed a considerable challenge. Unlike stars, spiral nebulae are extremely faint, and even the brightest, Andromeda, is only just visible to the naked eye. With little light available in the first place, and even less once it had passed through the prism of a spectroscope, Slipher had at first despaired of photographing a single spectrum. For two years he had tried everything he could think of, without success. Then, in late 1912, after the installation of a better camera, he at last obtained a spectrum for the Andromeda nebula. When he developed the plate he received a shock. Andromeda's light was shifted towards the blue end of the spectrum. The nebula was hurtling towards Earth at the alarming speed of 300 kilometres per second. Nothing like it had ever been seen before.

Over the next year Slipher photographed the spectra of another 14 nebulae, with varied results: all appeared to be moving, but in only one other case was the light shifted towards the blue end of the spectrum, the rest all had red-shifts. The implications were clear: Andromeda was an excep-tion. Most nebulae were receding from the Earth, and, in some cases, receding fast. Two were zooming away at a brisk 1,100 kilometres per second.

The discovery raised a fascinating question: what was caus-ing the nebulae to move so fast?

The theoretical framework that would eventually provide the answer to this question was laid down in Berlin during the First World War. In 1916 Einstein published his General Theory of Relativity, a groundbreaking work that tied up many of the loose ends of physics that had been lying around since Newton's time. In abstract terms, the theory describes the mathematical relationship between space, time, gravity and matter, but almost as soon as he completed it Einstein realised that it also offered the tantalising possibility of describing the actual size and nature of the universe itself. He immediately set to work drawing out the cosmological implications of his theory – a task he found so challenging that he told a friend it left him vulnerable 'to the danger of being confined in a madhouse'.

Ever since Newton, most physicists had considered that the universe was infinite, stretching for ever in all directions. But on the strength of his theory, Einstein rejected this belief in favour of a finite universe. When he worked through the equa-tions that described this new universe, however, he was sur-prised to find that it wasn't static, as he had imagined, but was either expanding or contracting. This was ridiculous, or so he thought. Although he didn't know much about astronomy,

Einstein felt certain that the stars weren't flying off into space or crashing down towards us. (At this point he hadn't heard about Slipher's red-shift measurements of spiral nebulae.) He therefore introduced an additional term to the equations, a fudge factor he called the 'cosmological constant', that would keep his theoretical universe static.

As the saying goes, there is more than one way to skin a cat, and before long, another physicist had derived a completely different solution to Einstein's equations. At first sight, the solution proposed by Willem de Sitter, the professor of astronomy at Leiden in Holland, appeared more elegant. Unlike Einstein's it was naturally static, with no need to add the extra term of a cosmological constant. (In science the simplest solution is normally considered the most beautiful and most likely to be correct.) However, this was only achieved at a cost. To remain stable de Sitter's model universe had to be empty, which – as Einstein was quick to point out – meant it couldn't contain any matter, not even a single star. In reply de Sitter argued that, as long as the amount of matter in the universe was small compared to its volume, his model would still be viable.

Intriguingly, de Sitter's universe also possessed a curious feature that wasn't shared by Einstein's: distant clocks would appear to run slower, and the wavelength of light emitted by faraway stars would appear to lengthen – in other words, display a red-shift. This wasn't because the stars were moving away; it was simply an impression produced by the particular properties of space and time in de Sitter's model. What was important, though, was that the predicted red-shift offered a way to check the theory. If de Sitter was right, then the red-shift of distant stars would be greater than that of nearby ones.

In the mid-1920s, following Hubble's discovery that spiral nebulae were in fact distant galaxies, a tiny handful of astronomers began searching the heavens to see if de Sitter's relationship between red-shift and distance held true. It was quite a challenge. As Slipher had found, accurate measurements of red-shift took skill and patience, while precise distances to nebulae and other remote celestial objects were even harder to obtain. There was also the added confusion that the Andromeda galaxy was hurtling in the opposite direction to that predicted by de Sitter's theory. (Though this could be explained by the fact that this nearby galaxy and our own were being drawn towards each other by their mutual gravitational attraction.) Despite the efforts of these astronomers, no one was able to find the evidence to back up de Sitter's theory.

Meanwhile, with the war over, and Einstein's general theory now common knowledge, other scientists were getting to grips with its implications. Among them was Alexander Friedmann, a Russian mathematician turned meteorologist working in Petrograd (St Petersburg). The son of a ballet-dancer father and piano-teacher mother, Friedmann had spent the war in the Russian volunteer aviation detachment, dropping bombs on Austrian troops. After the Russian revolution of 1917, he returned to Petrograd where, in the early 1920s, he first derived then solved the cosmological equations implied by Einstein's theory. Unlike both Einstein's and de Sitter's solutions, Friedmann's indicated an expanding universe.

When he read Friedmann's paper in the journal *Zeitschrift für Physik* in late 1922, Einstein wasn't impressed. He was so sure that Friedmann's proposal of an expanding universe was wrong that he wrote a letter to the journal warning them that the results 'appear suspicious', and that the solution 'does not satisfy the field equations'. Eight months later he was eating humble pie. Friedmann had written to him, argued his case, and won. Einstein's apology appeared the following May:

In my previous note I criticised the above-mentioned work. However, my criticism . . . was based on an error in calculation. I consider that Mr. Friedmann's results are correct and shed new light.

Years later, Friedmann's students at Petrograd would proudly recall the day their professor proved Einstein wrong. However, for some reason that nobody has ever satisfactorily explained, Friedmann's paper was soon forgotten. This may have been because Friedmann himself didn't live long enough to champion his own cause. In August 1925 he made a record-breaking high-altitude balloon ascent to make medical and meteorological observations. Shortly afterwards, still physically weakened by the experience, he caught typhoid. Within weeks he was dead. For the next five years, for all the world knew about it, his theory of an expanding universe might as well have died with him.

Meanwhile the question remained of why spiral nebulae were speeding through space. In the summer of 1928 the International Astronomical Union held a major meeting on nebulae in de Sitter's home town of Leiden, and Hubble travelled to Europe to chair the committee. Even though he had better facilities than anyone else to investigate the relationship between distance and red-shift, up to this point he had taken little interest in de Sitter's theories. However, something at the conference – possibly a conversation with de Sitter himself – now stirred him into action. When he arrived back in California that August, he called on Milton Humason (the assistant astronomer whose observations of cepheids Shapley had reportedly dismissed) and outlined his plan of attack.

Of all the observers he could have paired up with, Humason was undoubtedly the best. His patience, perseverance and attention to detail wrung images out of the 100-inch telescope his colleagues could only dream of. Yet this easygoing, round-faced Minnesotan had no academic training. He had started off on the mountain as a mule driver carting parts of the first telescopes to the summit. Some years later, after working as a relief night-assistant, his ability had been recognised and he had been taken on to the staff full-time. Hubble's programme would now push his skills – and his patience – to the limit.

The division of labour was straightforward. While Hubble worked out distances, searching for nebulae further and further away, Humason obtained their red-shift. As Hubble readily admitted, this latter task proved the greater physical challenge. The further they pushed out into the cosmos, the less light reached the photographic plate, which meant increasingly long exposures: it could take four nights to obtain a single spectrum. As winter approached, the mountain became bitterly cold. Inside the dome, heating was banned as the rising air would have created turbulence over the telescope, distorting the image they were trying to capture. 'Nobody who hasn't done it could ever realise how cold we were,' Humason recalled. 'It was impossible to keep warm.' For hours on end he would sit motionless, his eye glued to the eyepiece, keeping the guide-star on the cross-wires, while his thumb hovered over the button that advanced or retracted the telescope, ready to compensate for the slight flaws in the 100-inch tracking system. It took total concentration.

Hubble drove him on. 'I didn't feel much enthusiasm about these long exposures, but he kept at me and encouraged me ... When I got back from the mountain, he would come striding down the hall to ask what luck I had had, he was always so interested and eager about it.'

By January 1929 they had obtained spectra and distances for 24 galaxies ranging as far as the Virgo cluster, which Hubble had estimated as 6.5 million light-years away. This was too far away to spot cepheids, so Hubble had estimated its distance by measuring the brightness of its brightest stars. (He assumed that the brightest stars in any galaxy all emitted the same amount of light.) All the distant galaxies displayed red-shifts, zooming away from the Milky Way with increasing speeds. When Hubble plotted the velocity of each galaxy against its distance, he found that they all fell on a straight line. This was remarkable. A galaxy twice as far away as another appeared to be travelling twice as fast, while one three times as far away appeared to be travelling three times as fast, and so on.

Eager to confirm this result, Hubble and Humason forged on, obtaining distances and red-shifts to fainter and fainter nebulae. In 1931 they reached the Leo cluster, by Hubble's reckoning a staggering 105 million light-years away. Its speed of recession fairly took the breath away – 19,700 kilometres per second. Even better, the red-shift to distance ratio was bang in line with all their previous results. The scientific world was agog. There appeared to be an underlying reason why the galaxies were zooming away, but no one was sure what it was.

What quickly became clear was that the simple linear relationship Hubble had found was not the one predicted by de Sitter. Instead of vindicating de Sitter's model of the universe, Hubble's discovery knocked it on the head. The same was true for Einstein's model, which predicted that the nebulae should be stationary. Having demolished two theories, Hubble was left with no theoretical explanation for his discovery. Astronomy was faced with a theoretical vacuum, a phenomenon with no explanation. Theoreticians around the world – unaware of Friedmann's paper – began searching for a solution. They didn't have long to wait.

During a discussion on the subject at a meeting of the Royal Astronomical Society in London in January 1930, the Cambridge astrophysicist Arthur Eddington made a chance remark. 'One puzzling question,' he said, referring to Einstein and de Sitter's model universes, 'is why there should be only two solutions. I suppose the trouble is that people look for

Expanding Horizons

Spectra showing increase in red-shift with increasing distance.

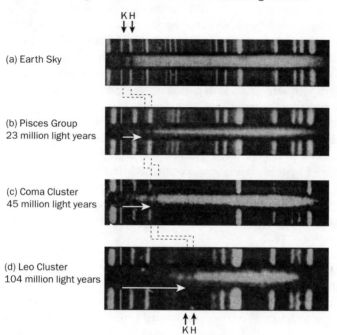

Hubble's distances, on the left, correspond to Humason's spectra on the right. The central horizontal band in each image is the spectrum of a distant galaxy, while the barcode-like image it is superimposed over is a reference spectrum. The white arrows indicate the amount of red-shift.

a) The two prominent dark lines (known as H and K lines) are caused by calcium absorption.

b) Galaxy NGC 385; apparent velocity 4,900 km/sec.

c) Galaxy NGC 4884; apparent velocity 6,700 km/sec.

d) Brightest galaxy in the Leo cluster; apparent velocity 19,700 km/sec static solutions.' A few weeks later, much to his surprise, he received a letter from Belgium drawing his attention to a third solution, one that wasn't static but dynamic. In all essentials the paper was identical to Friedmann's – only it wasn't Friedmann's; it had been written three years earlier by Georges Lemaître, a former pupil of Eddington's who was now professor of astronomy at Louvain University in Belgium. Working alone, and completely unaware of Friedmann's earlier paper, Lemaître had independently reached the same conclusion – the universe was expanding.

With Friedmann dead and his theory temporarily forgotten, it was Lemaître who first provided the world with a satisfactory explanation of Hubble's results. In doing so, he not only proved Einstein wrong and confirmed that the universe was growing bigger by the day, he also brought science to bear on the fundamental religious question that had triggered the search for the beginning of time in the first place – the creation of the universe.

If you should ever chance to flick through an album of those obligatory group photographs taken at astronomy conferences in the 1920s and 30s you will easily spot Lemaître. As an ordained Catholic priest, his black soutane and dog collar leap out from the grey, serried ranks of besuited astronomers. At just nine years old, Lemaître had made up his mind to become both a scientist and a priest. 'There were two ways of arriving at the truth,' he later told a journalist. 'I decided to follow them both.' After leaving college he had begun training for a career in the mining industry but, when Germany invaded Belgium in 1914, had abandoned this to fight for his country. In the early years of the war he had been involved in some bitter house-to-house fighting and had later witnessed the appalling consequences of the first use of chlorine gas, which left a lasting impression. When the conflict ended he had studied for a doctorate in mathematics, then trained for the

Catholic Church. After travelling in Canada and the United States, working alongside a number of astronomers, he had returned to Louvain in Belgium, where in 1927 he had tackled and solved the cosmological equations derived from Einstein's theory.

Like Friedmann, Lemaître had found that his solution indicated an expanding universe, and – just like Friedmann's solution – his own also lay forgotten for several years. In part this was because it was published in a little-read Belgian journal, but, once again, Einstein had failed to appreciate the significance of someone else's work. When Lemaître had shown him his paper at the Solvay Conference in Brussels towards the end of 1927, he had dismissed it with an uncharacteristic snub. 'Your calculations are correct, but your physics is abominable.' Even when Lemaître told him about the speeding nebulae that Slipher had found, he showed little interest, and Lemaître had come away with the impression that he wasn't well informed on the latest developments in astronomy. Twice Einstein had been offered the opportunity to recognise his mistake; twice he had failed. His mind was set on a static, unchanging universe. Later, once he had seen the beauty of Friedmann and Lemaître's solutions, he would change his mind, calling the cosmological constant (the 'fudge factor' he had invented to hold the universe static) the 'biggest blunder' of his life.

As a consequence of Einstein's negative assessment, Lemaître's paper lay unnoticed for three years, until, in February 1930, he happened to read about Eddington's chance remark at the Royal Astronomical Society. Realising he had the answer that everyone was looking for, he wrote to Eddington, drawing attention to his paper. Within a year Friedmann's solutions were also rediscovered, and, together with Lemaître's, provided the explanation for Hubble's high-speed galaxies. The galaxies weren't flying through space as nearly everyone had assumed; space itself was expanding. The universe was getting bigger by the day.

Scientists were stunned. Like Copernicus's revelation that the Earth revolved around the Sun, Hubble's discovery completely changed humanity's view of the world. If the universe had been found to be static and unchanging, then, as far as science was concerned, it might as well have been an eternal universe; for if there was no way of measuring any development, there was no way of probing its past. But Hubble's discovery that the universe was expanding presented science with an exciting proposition. An expanding universe meant a changing universe – a universe with a history that, one day, they might possibly measure. But first they had to make sure of their facts. If the universe was expanding, the next question was: what was it expanding out of?

Friedmann had suggested that it had all originated out of a single point, but Lemaître couldn't go quite that far. He knew that one of the fundamental tenets of physics was the idea of the conservation of matter: although matter could be transformed into energy and vice-versa, it couldn't just be created or destroyed. Whatever the original universe looked like, it must have contained all the mass and energy of the present one. With this idea in the back of his mind, Lemaître began forming a theory of how the universe had evolved. His inspiration came from radioactive decay. It was well known that heavy atoms, such as uranium, decayed into lighter ones, but perhaps, he reasoned, uranium and the other heavy atoms had themselves been produced by even heavier atoms. Following this train of thought back to its logical conclusion he arrived at the notion that the universe had sprung from a single, supermassive 'primordial atom', an idea he first expressed in a letter to *Nature* in May 1931: '. . . we could conceive the beginning of the universe in the form of a unique atom, the atomic weight of which is the total mass of the universe. This highly unstable atom would divide in smaller and smaller atoms by a kind of super-radioactive process . . . until our low atomic number atoms allowed life to be possible.'

This, then, was the picture Lemaître had in his head when he arrived at Pasadena in December 1932 to present his ideas to Einstein, Hubble and a host of interested scientists. Standing in front of one of the most discerning audiences any scientist has ever had to face, his usual shyness temporarily forgotten, Lemaître launched into his explanation. At the origin, all the mass of the universe existed in the form of a unique primeval atom, the radius of which, although not strictly zero, was relatively small. This primeval atom had spontaneously disintegrated into smaller 'atomic stars' – dense superatoms, each weighing as much as a single star. When these in turn had also disintegrated, they created the matter of the present-day universe. 'In the beginning of everything we had fireworks of unimaginable beauty,' he enthused. 'Then there was the explosion followed by the filling of the heavens with smoke. We come too late to do more than visualize the splendor of creation's birthday!' When Lemaître finished, Einstein rose to his feet. Six years earlier he had called Lemaître's physics 'abominable'. Now he praised it to the rafters. 'This is the most beautiful and satisfactory explanation of creation to which I have ever listened,' he gushed. Lemaître blushed with delight.

The picture painted of a single atom decaying is very different from the picture of the beginning of the universe we have today; nevertheless the lecture marked a fundamental turning point in science. Lemaître swept away the old idea of a static, eternal universe and replaced it with a new image, that of a dynamic, evolving universe that had changed over time; a universe that had a beginning: 'a day without yesterday', as he memorably called it. The murky waters of metaphysical speculation that had shrouded the moment of creation for

Albert Einstein and Georges Lemaître in Pasadena at the time of Lemaître's talk.

centuries were beginning to clear. For the first time, scientists glimpsed the real possibility of finding the moment the universe began.

While many scientists harboured doubts about Lemaître's explanation of how the universe began, nearly all accepted his notion that the universe was expanding. They also realised that if the rate of this expansion was known, then it should be possible to work backwards and discover how long it had taken to reach its present size.

Attention now focused on Hubble's results. In a 1931 paper jointly written with Humason he put the rate at which the universe was expanding (now known as the Hubble constant) at 558 kilometres per second per megaparsec. This meant that two galaxies separated by a megaparsec (a distance of about 300 million light-years) were moving apart from each other at 558 kilometres per second. On such a large scale, this high speed makes the expansion appear phenomenally fast. However, on a scale that is easier to visualise we can see that it is really quite slow. For instance, two points in space separated by a distance of 1,000 kilometres would be moving apart from each other at a rate of just over half a millimetre per year. The reason that distant galaxies appear to move faster than nearer ones is because there is more space between them and us, and each part of that space is expanding.

Eddington used to explain this to his students with a simple analogy. He would ask them to imagine what would happen if the size of the classroom doubled. As the room grew bigger, two adjoining desks would move perhaps a few inches further apart, while two desks on either side of the class would move several yards apart. A student sitting at any desk would see his close neighbours inching away slowly, while the desks on the far side of the room would appear to travel away quickly. The same thing, on a much larger scale, was happening to the galaxies in space.

Once Hubble and Humason had determined the rate of expansion, it was a simple matter to find the universe's age. Knowing how fast the galaxies were moving away from each other, and assuming that they had always moved apart at this rate, astronomers worked backwards to find the moment they had all emerged from a single point. When they did this, in 1931, they found that the universe was 1.8 billion years old.

Almost as soon as they had determined this new figure, however, astronomers realised that something was seriously wrong with it. In 1927, using radioactive dating, Holmes had estimated the Earth's age as somewhere between 1.5 and 3 billion years, while in 1929, Rutherford had put it at 3.4 billion years. The evidence of astronomy conflicted with that of geology. According to the astronomers, the universe was younger than the objects it contained.

But worse was to come. The age of 1.8 billion years had been derived on the assumption that the universe had always expanded at the same constant rate. However, Einstein and most astronomers knew that this was unlikely to be true. They believed that the expansion was slowing down. The universe was full of matter in the form of gas, stars and galaxies, and every piece of this matter exerted a gravitational pull on every other piece. The net effect was an inward pull, slowing the rate of expansion.

Another way of explaining this is by comparing the expansion of the universe to a stone thrown into the air. As the stone leaves the hand it rises quickly, but from then on, the force of gravity slows its ascent. Likewise, the Big Bang caused the universe to expand quickly at first, but the gravitational pull of all the mass within it has since caused the rate of expansion to slow down. It was even possible, some suggested, that the gravitational attraction of all the matter in the universe might one day cause it to collapse in on itself.

The result of this realisation, that the universe was probably

Expanding Horizons

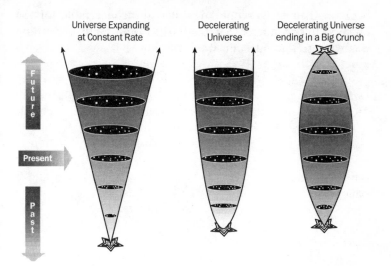

Universe Expanding at Constant Rate

Decelerating Universe

Decelerating Universe ending in a Big Crunch

Future

Present

Past

slowing down, reduced the estimate of the universe's age. Because the universe had expanded faster in the past, it had reached its present size in less time than the present rate of expansion suggested. Instead of being 1.8 billion years old, it would be slightly but significantly younger.

This exacerbated the conflict with geology. In the coming years, estimates based on radiometric dating went from strength to strength as confidence in the method continued to grow. Announcements from geologists in the 1930s and early 1940s put the Earth's age at between 3 and 4 billion years old. At the same time, in the absence of new data, the estimate for the universe's age based on a constant rate of expansion sat at 1.8 billion years – though reason now suggested it should be younger. Faced with these conflicting results, the expanding universe theory, just for a moment, began to falter. Everyone knew it was inconceivable for the universe to be younger than the Earth. Somewhere, something was wrong. The question was, what?

14. Chasing a Rainbow

On the morning of 7 December 1941, Japanese planes launched a ferocious attack on the United States Pacific Fleet as it lay at anchor in Pearl Harbor, Hawaii. The following day Congress declared war on Japan, and the American nation braced itself for the struggle ahead. 'This is rather an important thing,' Hubble told his wife, as they stood in the driveway of their Pasadena home and watched the lights of Los Angeles flicker off in the valley below. The government had just ordered the first of many blackouts, and that evening the entire west coast of America, as far inland as Las Vegas, was plunged into darkness.

Two thousand miles to the east, in the little town of Oxford, Ohio, fifteen-year-old Allan Sandage peered up at the stars through his back-yard telescope as he did most nights of the week. An only child, Sandage was overwhelmed by the wonder of the world around him. It filled him with a deep religious instinct, and on Sunday mornings, while his parents slept in, he would slip out alone to attend the local church service. But he also yearned to understand the workings of nature, how it functioned, what made it tick. At the age of nine he had persuaded his father to buy him a small, one-and-a-half-inch, telescope, learnt to develop his own plates, and in a prescient display of quiet determination, kept a log of the number of sunspots he saw each day for five years. When he read *The*

Realm of the Nebulae, Hubble's account of his discoveries, he realised astronomy 'had gotten into the seed of my soul'. There was no turning back.

That last summer before the war, he and his father had motored west to California, where they had made a special detour to Mount Wilson. The drive up the precipitous winding road to the observatory had terrified his father, who had rarely left the flatlands of Ohio, but Sandage had been in his element. He stared in awe at the 100-inch telescope with which Hubble had discovered the expanding universe, and learnt with growing excitement about an enormous new telescope being built on Mount Palomar, some ninety miles to the south-east. Twice as powerful as the 100-inch, the new colossus would trawl the extreme depths of the cosmos, revealing galaxies never seen before. To work with this new giant became his burning ambition. By the time of the Japanese attack on Pearl Harbor, he had resolved to go to graduate school and become an astronomer.

With the United States' entry into the Second World War, any immediate hopes of resolving the discrepancy between the age of the Earth and the age of the universe vanished. Astronomers were needed elsewhere. Anyone capable of determining the orbit of a planet, or the path of a comet, could equally well calculate trajectories of bombs and mortars. A steady stream of scientists, including Hubble, left Mount Wilson to work for the military.

One of the few astronomers who remained on the mountain was Walter Baade, a warm-hearted German émigré, ineligible for war work because of his nationality. A meticulous observer, Baade soon turned the situation to his advantage. With the nearby city of Los Angeles blacked out, and almost as much time on the 100-inch telescope as he could use, he not only made new discoveries, he rapidly became the most accomplished observer in America. After the war, when the new

200-inch telescope finally opened on Mount Palomar, it was Baade who resolved the conundrum of why the universe was younger than the earth.

Baade's specialty was RR Lyrae stars – variable stars like cepheids, but many times fainter. Like cepheids, the distances to the nearest RR Lyrae stars had been measured using statistical parallax and their absolute luminosity determined. This meant that they could be used as distance indicators, although because of their faintness, their use was limited. Baade had observed them in our galaxy and the neighbouring Magellanic Clouds with the 100-inch telescope, but they were too faint to be seen in any other galaxy. According to Baade's calculations, however, if Hubble's distance to Andromeda was correct, he would just be able to resolve RR Lyrae stars in that galaxy with the new 200-inch telescope on Palomar. When the telescope opened in 1949, they were naturally the first objects he looked for.

When he trained the telescope onto Andromeda and ran off the necessary exposure, however, the RR Lyrae stars were nowhere to be seen. After weeks of trying and much deliberation, he realised that there was nothing wrong with his calculation, or the optics of the telescope; the whole galaxy was simply much further away than Hubble had calculated. Some idea of how much further was revealed when he studied his new plates. Bright red giant stars, that Baade knew to be four times brighter than RR Lyrae stars, could just be discerned, displaying the same apparent brightness he had expected the RR Lyrae stars to show. This meant that all the stars in Andromeda were four times fainter than they should have been if the galaxy was at the distance Hubble indicated, or – as the apparent brightness of a star decreases in proportion to the square of its distance – the galaxy was twice as far away as originally thought.

The reason for Hubble's error soon emerged. He had based

his estimate on Shapley's cepheid distance scale which, it now emerged, was wrong. While the distance to the RR Lyrae stars had been determined relatively accurately, dust in the galaxy had affected the calibration of the cepheid scale, with the result that it was out by about a factor of two. The distances to all the galaxies Hubble and Humason had found now had to be doubled.

Baade's discovery was welcomed with open arms. By doubling the distance to Hubble's galaxies, he not only doubled the size of the universe, he doubled its age. At a conference in Rome in 1952 Baade announced that the period of time which had elapsed since the universe began should be increased to 3.6 billion years. At last the universe was older than the Earth.

But this wasn't the end of the story. Finding the expansion rate of the universe, as astronomers soon discovered, was like chasing a rainbow – however hard they tried to pin down its value, the true rate always appeared one step away. But like eighteenth-century members of the French *Académie* setting out to measure the circumference of the Earth, at least they knew they were chasing something that could be measured; by the 1940s there was no doubt that the universe was expanding. Whether it truly had a beginning or not, however, was still open to question. Lemaître had failed to produce any solid evidence in support of his idea that everything was born out of an endlessly splitting atom. Indeed, apart from the observed expansion, there were no tangible clues that the universe had a moment of birth at all.

All that changed, however, when George Gamow, an eccentric Russian-born physicist working in Washington, DC, brought his vivid imagination to bear on the problem. In an exciting series of papers published in the late 1940s, he and two colleagues, Ralph Alpher and Robert Herman, explored the dramatic events that took place in the first few seconds of the universe, and, in doing so, laid the foundations for the modern theory of the Big Bang.

George Gamow, or 'Joe' as he was popularly known, was a fun-loving, irreverent practical joker; a 'swashbuckling scientist' sparkling with bright ideas who never took himself or his work too seriously – much to the irritation of many of his colleagues. At university in Leningrad, Gamow had studied cosmology under Friedmann, before winning international recognition in the 1920s by explaining how alpha particles 'tunnel' their way out from the nuclei of radioactive elements. After fleeing the Soviet Union in 1933, he settled in America, where his boundless energy, atrocious spelling and childlike sense of fun became as celebrated among his new colleagues as his continued accomplishments in the field of physics. 'I took the liberty,' he wrote to one collaborator, 'of completely rewriding [sic] your letter to *Nature*, the original version was too serious.' By the 1940s he had also become well known to the American public as the author of the *Mr Tompkins* series of popular science books. But the lasting achievement of this boundless creativity was his work on the physics of the early universe.

From the start, his approach to the events that occurred in the first seconds of the universe's life was radically different to Lemaître's. Instead of starting the universe with a single primeval atom splitting into smaller and smaller atoms, Gamow began with a huge number of tiny particles and built up atoms through collisions between them. His initial universe was just as compact, just as dense as Lemaître's, but instead of a single, solid entity, it consisted of a thick, intensely hot 'soup' of neutrons – the small elementary particles found in the nuclei of atoms. These were his building blocks. When neutrons collided with one another in the 'soup', they sometimes linked together, building up, through further collisions, larger and larger 'neutral complexes', the nuclei of future atoms.

The major difficulty Gamow faced was that there were few clues as to what had actually happened all those years ago. Time had drawn a veil over the past; there was virtually no evidence against which he could test his theory. But one tiny piece of information caught his attention. In 1937, the Swiss-Norwegian geochemist Victor Goldschmidt had published a table of the relative abundance of the various elements. It showed that heavy elements, such as iron and lead, were scarce in the universe, while the two lightest elements, hydrogen and helium, were exceedingly abundant. It wasn't much to go on, but it was enough. Gamow realised that light elements, comprising just a few neutrons, could be built comparatively quickly with just a few collisions between neutrons, while to build a single atom of a heavy element, such as iron or lead, took a large number of collisions and a long time.

Here was a possible explanation for the relative abundance of the various elements. As the universe expanded in the first few seconds of its life, it would have rarefied and quickly cooled. This expansion, Gamow believed, took place so rapidly that the hot, dense conditions necessary for nuclear reactions to take place would have existed for only a limited time. As neutrons combined to form nuclei one at a time, he guessed that this period was not long enough for more than a tiny fraction of neutrons to build up into the large nuclei necessary to make heavy elements. Hence their scarcity.

At this stage Gamow's idea was little more than a rough sketch. To turn it into a fully-fledged theory he would have to work out the reactions of the particles in detail. It was a ghastly thought. Details had never been his forte and the prospect of endless calculations filled him with dread. He therefore suggested to Ralph Alpher, one of his graduate students, that it might be an interesting subject for his PhD.

Alpher readily agreed to Gamow's suggestion, and soon the two of them could be seen, on late afternoons, thrashing out the problems over drinks at the Little Vienna, a dimly lit downtown bar in Washington, DC. Soon, Alpher began discussing the project with his close colleague Robert Herman, a specialist in spectroscopy, and before long he joined the project as the third member of the team. Together the 'Three Musketeers', as Gamow jokingly called them, set out to put some flesh on the skeleton of his theory.

The first advance came from Alpher. After extensive calculations he was able to show that the distribution of the elements predicted by the theory almost exactly matched their distribution in the real world. Hydrogen and helium would have been produced in vast amounts, but there would have been fewer and fewer of the successively heavier elements. Plotted on a graph, the predicted abundance of each element showed an uncanny similarity to their actual abundance. If Alpher's calculations were right, and he had every reason to believe they were, then all the atoms in the universe had been synthesised at the dawn of time.

The paper in which Alpher and Gamow announced these results in 1948 is now famous. By providing a solid theoretical explanation for the observed abundance of the different elements, it demonstrated that particle physics was a valid tool for probing the workings of the early cosmos. More importantly, it convinced many people that the universe was truly created at a single moment in time. In many ways, the paper heralded a new era in our understanding of the cosmos.

There was, however, one nagging doubt, a potential flaw in their theory that plagued their dreams. It became known as the mass-gap problem. The difficulty emerged when they examined *how* the elements were made in the neutron 'soup'. Each element had to be built up one neutron at a time, a bit like building a chain, link by link. After each addition, the resulting proto-atom had to be stable, before the next neutron could be added. Unfortunately, in nature there are no stable elements with atomic masses of 5 or 8. At these positions – where physicists would like a link to the next element – there is only a yawning chasm, a 'mass-gap' that somehow has to be bridged. However hard they tried, Gamow and his team couldn't construct elements beyond atomic weight four. Hydrogen, deuterium, tritium and helium all surrendered to their equations, but there creation stopped.

For a time Gamow lived in hope that a solution would be found. As late as 1952 he believed it was simply a matter of 'considering the thermonuclear reactions in somewhat more detail'. But no matter how hard he tried, neither he nor anyone else could find a solution. It later turned out that they were looking in the wrong place. In 1957 a team of four astrophysicists, Margaret and Geoffrey Burbidge, William Fowler and Fred Hoyle, found that the heavier elements were produced by the high pressures and temperatures inside stars.

Their fascinating explanation still stands today. In the millions of years that immediately followed the birth of the universe, the giant clouds of hydrogen and helium gas produced by the initial explosion slowly collapsed under the pull of gravity to form stars. Carbon, oxygen, nitrogen, calcium, iron – most of the elements from which we are made – were produced inside these stellar furnaces, then blown out across space millions of years later when the stars ran out of fuel and exploded. Pulled together by gravity, the debris from these explosions formed new stars, and just occasionally planets like the Earth. By the time all this was discovered, however, Gamow had long since lost interest in the problem.

In the mid-1950s the 'Three Musketeers' split up to go their separate ways. Gamow, ever the maverick, spun off into the field of biology, while Alpher and Herman left the academic world to work in industry. Before they parted, though, the team planted a seed for the future. Tucked away in a paper published in 1948, Alpher and Herman left a telling prediction. According to their calculations, the remnants of light and X-rays released in the initial fireball at the birth of the universe should still be visible. In the billions of years since the explosion, the expanding universe would have red-shifted this radiation to longer and longer wavelengths until, no longer visible as light, it would exist only in the form of microwaves. These would pervade the universe as a background radiation with an energy corresponding to a temperature of 5 K. If it could be detected, it would provide confirmation that the universe was born in a single fiery blast. Because they believed the radiation would be too weak to measure, Alpher and Herman didn't pursue the idea. Instead, for the next seventeen years, their prediction remained buried in the journals.

Meanwhile, back in California, twenty-two-year-old Allan Sandage was halfway towards his dream of becoming an astronomer. In 1948 he had won a coveted place on the first astronomy PhD programme run at the California Institute of Technology, 'Caltech' for short, in Pasadena. The course was tough, but he didn't care; he was close to the action, a mere stone's throw away from the offices of the Mount Wilson Observatory.

Even so, his big break came sooner than he could have dreamed. One day in May 1949 Jesse Greenstein, his tutor, walked into the room with some news. Edwin Hubble had phoned asking for an assistant, and Greenstein had suggested his name. 'Why don't you go up to Santa Barbara Street,' he said, 'and see what Mr Hubble wants.'

Sandage was so overawed at meeting the man he regarded as the greatest astronomer in 400 years, that decades later he was unable to recall a single detail of that first encounter. What quickly became clear was that Hubble wanted an assistant: initially to carry out the tedious work of counting galaxies on hundreds of photographic plates, but later to help operate the telescope at Palomar. After a short probation, Hubble took Sandage under his wing, employed him as his observing assistant, and set him on the course that was to occupy his whole career: the search for the age of the universe.

Ever since Baade's discovery of the problem with the cepheid calibration, Hubble had known that there was something seriously wrong with his value for the rate of expansion of the universe. Early results from the Palomar telescope had revealed that many of the observations he had made with the 100-inch telescope all those years ago were unreliable. He now realised that the whole distance scale on which the rate of expansion was based was riddled with errors.

Essentially the problem came down to the difficulty of measuring the distances to remote galaxies accurately. And only remote galaxies would do. The rate of expansion of the universe, or Hubble constant, can't be determined from nearby galaxies, as the gravitational attraction between them and the Milky Way counters the effect of the expansion; they are receding from us more slowly than the universe's true expansion rate. To this day, however, measuring the distances to remote galaxies is fraught with difficulty. While the speed, or red-shift, of a remote galaxy can be determined from a single direct measurement of its spectrum, the distance to the galaxy has to be built up step by step. When Hubble first pushed out into the cosmos, he had been forced to adopt a different method of measurement for each stage of the journey. For the first hop, out to the six closest galaxies, he had relied on observations of cepheids. Beyond this, cepheids were too faint to be seen, so he had turned to the brightest stars in galaxies, assuming they were all of the same luminosity. Further out still, when he could no longer resolve individual stars, he had resorted to comparing the brightnesses of whole galaxies, assuming that these too could be used as 'standard candles'. It was on this string of rickety assumptions that he had based his initial value for the rate of expansion, but now he realised that his data were wrong, there was nothing for it but to begin the huge project again, to start from scratch, this time with the 200-inch telescope. With Sandage's help, he planned to double-check every observation, reassess every assumption, to do whatever it took to find the true rate of the expansion.

With the arrival of the idea that the world had all sprung from a single point at a single moment in time, the figure of God re-entered the scientific debate. Ever since the time of Descartes, rationalist thinkers had clashed with Christian theologians over the question of the world's age. But now, with Lemaître's and Gamow's assertion that there had truly been a beginning to the universe just as the Bible said there had, science and religion appeared to be singing from the same hymn sheet. Although Lemaître always kept his scientific and religious views separate, never suggesting that the explosion of the primeval atom was synonymous with the biblical act of creation, many commentators immediately linked the two events. In the absence of a scientific explanation, they saw God as the instigator of the explosion that had begun the universe, the hand that had lit the fuse. This apparent link with religion provoked extreme reactions.

The most hostile came from within the Soviet Union. Ever since the Bolshevik revolution of 1917, the Soviet government had condemned religion as superstition, and, in an attempt to impose its own doctrine of materialist atheism on the people, had violently repressed the Orthodox Church. Any philosophy that smacked of religion was considered suspect, and Lemaître's theory of an exploding universe with its echoes of biblical creation was no exception. The fact that a Catholic priest had first proposed the idea practically guaranteed its condemnation. Those suspected of promoting it were branded

'Lemaître's agents', while Lemaître himself and his Western supporters were denounced as 'falsifiers of science' who wanted 'to revive the fairy tale of the origin of the world from nothing'.

Given the brutal nature of Stalin's oppressive regime, it's not surprising that nearly all Soviet astronomers publicly rejected the exploding universe theory. Two scientists who had earlier supported Friedmann's ideas of an expanding universe, Matvei Bronstein and Vsevolod Frederiks, were arrested in the 1937 purge of physicists. Bronstein was falsely convicted of being a foreign spy, and was shot by firing squad the following year, while Frederiks was sent to a string of concentration camps. He died in transit between two camps six years later. With their deaths, Soviet research into the origin of the universe practically ceased.

In Europe, a more rational objection to the exploding universe theory emerged in England, though this too was born from atheism. It came from the English physicist Fred Hoyle, an irrepressible Yorkshireman with a deep-seated distrust of religion. Hoyle accepted that the universe was expanding, but found the whole idea of a single moment of creation totally improbable. If the universe had been thrown apart at the enormous speeds indicated by Hubble, he suggested, then it would have been impossible for gravity to pull the matter in the universe together to form galaxies. 'The two concepts of explosion and condensation are obviously contradictory,' he insisted. 'If you postulate an explosion of sufficient violence to explain the expansion of the Universe, [then] condensations looking at all like the galaxies could never have formed.'

Together with two other Cambridge physicists, Hermann Bondi and Thomas Gold, Hoyle proposed an alternative explanation, known as the 'Steady-State' theory. This asserted that, far from being born in a giant explosion, the universe was eternal and had always existed. To explain the expansion,

Hoyle and his colleagues suggested that matter is being continuously created in the space between galaxies. As the universe expands, the galaxies move apart, but the continuous creation of matter forms new galaxies in the gaps left behind, the overall effect being that the universe maintains a constant density and always looks the same: it exists in a 'steady state'. Only a tiny amount of matter was required to account for the observed rate of expansion: according to Hoyle, no more than 'one atom in the course of about a year in a volume equal to St Paul's Cathedral'.

The response to the Steady-State theory was mixed. In America, none of the Pasadena astronomers took it seriously. Their observations of elliptical galaxies revealed that they all looked the same age. With no sign of the young galaxies Steady State predicted, they discounted the theory immediately. European astronomers were less certain, and for a brief period in the 1950s and early 1960s Hoyle's Steady State enjoyed a modicum of support.

As the arguments raged back and forth, however, there was one thing both sides agreed on. In the spring of 1949, the British Broadcasting Corporation asked Hoyle to write and present a series of radio talks on the nature of the universe. Casting around for a suitable term with which to describe his opponents' theory, Hoyle coined the expression 'Big Bang'. It was meant as a term of derision, but much to Hoyle's amazement the name stuck.

Of all the reactions to the Big Bang theory, the most surprising came from the Roman Catholic Church. In the early 1950s Pope Pius XII, a scholarly but solitary man, became convinced that the theory provided scientific confirmation of the events described in Genesis. On 22 November 1951 he gave the theory a ringing endorsement in an address to the Pontifical Academy of Sciences:

. . . it would seem that present-day science, with one sweep back across the centuries, has succeeded in bearing witness to the august instant of the primordial *Fiat Lux* [Let there be Light], when, along with matter, there burst forth from nothing a sea of light and radiation, and the elements split and churned and formed into millions of galaxies.

Such confidence was premature. In scientific terms the Big Bang theory was hardly proven; many scientists strongly disagreed with it, and apart from Hubble's observations and Gamow's speculations, there was little evidence to back it up. But the Pope went even further:

Thus, with that concreteness which is characteristic of physical proofs, [modern science] has confirmed the contingency of the universe and also the well-founded deduction as to the epoch when the world came forth from the hands of the Creator. Hence, creation took place. We say: therefore, there is a Creator. Therefore, God exists!

News of the speech shot around the world. '. . . till I read this morning's paper,' a friend wrote to Hubble, 'I had not dreamed that the Pope would have to fall back on you for proof of the existence of God. This ought to qualify you, in due course, for sainthood.'

While some were amused, others were horrified. None more than Lemaître. He knew that the Big Bang theory was still little more than a hypothesis, and that hypotheses, like spring fashions, came and went with the season. It was crazy to rest one's faith on a theory that, although in vogue today, could be overturned by a better idea tomorrow. Unfortunately that was exactly what Pius XII had done. By asserting that God must exist because the universe started in a 'Big Bang', he had practically staked the Catholic faith on the Big Bang theory. A concerned Lemaître, together with the Vatican's science adviser, visited Pius XII and warned him that his approach was in danger of backfiring. If the Big Bang theory turned out to be wrong, he warned, it could only damage the Church. The Pope heeded his advice, and never raised the issue again.

The 1950s were a boom era for American science, a time of optimism and technological advance. Atomic power, jet planes, coloured plastics, tail-finned cars, television for the masses – it seemed there were no bounds to its achievements. The rapid strides in technology made during the war, not least the development of the atomic bomb, had demonstrated science's power to politicians and industrialists alike, and with the arrival of the Cold War they poured money into scientific research as never before. If scientists were the new pioneers, then space was the new frontier to be explored. Everything and everyone was looking up. Giant radio telescopes tracked the heavens; rockets blasted off from the New Mexico desert on a relentless programme of test flights; and a wave of science fiction movies played the drive-in cinemas. In homes across the country the talk was of life on other planets, and travelling into space. It was no longer just astronomers but the whole nation who turned their thoughts towards the stars.

Amidst all this excitement, Hubble and Sandage continued to chip away at the intergalactic distance scale, seeking the evidence that would provide the true age of the universe. At first progress was slow. Following a heart attack in the summer of 1949, Hubble's doctors had banned him from the mountain, so for a time Sandage had travelled up to Mount Palomar alone. Once Hubble was allowed to observe again, things picked up. They worked together as a team, re-photographing the galaxies on which Hubble had based his original distance scale, taking the first steps towards a new age for the universe.

Then in September 1953 Hubble suffered a cerebral thrombosis on his way home from work and died in the driveway of his home. 'It was such an incredible shock,' said Sandage. 'I walked out the door of Santa Barbara Street and walked around Pasadena by myself for two hours.' The blow was only compounded by the decision of Hubble's wife, Grace, to have the body cremated and to bury the ashes quietly in a secret location without telling his colleagues. There was no funeral service, no public ceremony, no announcement. It was as if Hubble had just disappeared.

But his programme remained. It was monstrous, fifteen or twenty years' work, maybe more. 'I felt a tremendous responsibility to carry on with the distance-scale,' recalled Sandage. 'He had started that, and I was the observer and I knew every step of the process that he had laid out . . . I knew at the time it was going to take that long. So, I said to myself, "This is what I have to do." ' At twenty-eight, Sandage, the youngest astronomer at the observatory, had the 200-inch telescope and the whole of Hubble's project to himself. The universe lay in his hands.

In the same month that Hubble died, the final chapter closed on the quest to find the age of the Earth. In 1953, the long search that had frustrated generations of scientists from Halley to Buffon, Kelvin to Rutherford, was finally resolved by an unprepossessing thirty-one-year-old geochemist from Des Moines in Ohio – Clair Patterson.

Radioactive dating had come on in leaps and bounds since Rutherford's day, and Patterson, a young researcher working at Caltech in Pasadena, was ideally placed to exploit the latest developments. One of the main discoveries in the preceding years was the realisation that there were two types of uranium, each of which decayed at different speeds. These two isotopes, as they were known, decayed into two separate isotopes of lead, and by the beginning of 1953 the two separate rates at which they decayed had been determined with great precision. This was a great leap forward; instead of one radioactive clock, geologists now had two. Each could act as a check on the other. However, before they could find the age of the Earth, geologists still needed one vital piece of information – the amount of each isotope of lead it contained at the time it formed.

If Boltwood's assertion that all the lead on the Earth had come from the radioactive decay of uranium was true, then it would be relatively easy to find the planet's age. The more lead the Earth contained, the older it must be. The process would be similar to looking at an hourglass where sand flows from the upper chamber to the lower one. If the upper chamber represents uranium, and the lower chamber represents lead, then as time passes, the upper chamber depletes and the lower one fills. Just as at any instant you can tell how long has passed by comparing the amount of sand in both halves of the hourglass, then you could find the age of the world just by measuring the relative amounts of uranium and lead.

In reality, however, not all the lead on the Earth was likely to have come from uranium. Some was almost certainly present when the Earth formed. It was as if the bottom half of the hourglass already had some sand in it when it started running. Without knowing how much lead already existed when the Earth formed, it was impossible to find its age. As Patterson succinctly put it:

> If we only knew what the isotopic composition of primordial lead was in the Earth at the time it formed, we could take that number and stick it into this marvelous equation that the atomic physicists had worked out. And you could turn the crank and blip – out would come the age of the Earth.

Fortunately Harrison Brown, Patterson's doctoral supervisor, had already thought of a way to find the primordial lead. The place to look was in meteorites. Brown knew that the components of the solar system – the Sun, planets, moons and asteroids – had all coalesced from one huge swirling disc of dust and gas at the same time. Meteorites, too, originated in this era, and Brown suggested that within one particular type – iron meteorites – lay a record of the lead ratios at the time the solar system formed. Although largely made up of iron, iron meteorites also contained tiny traces of lead, but – most importantly – they contained no uranium. None of the lead in the meteorite, therefore, could have come from radioactive decay: it was all primordial, and its isotopic composition was the composition of lead in the Earth at the time it formed. Patterson later recalled how Brown put the proposal:

So Brown said, 'Pat . . . You just go in and get an iron meteorite – I'll get it for you. We'll get the lead out of the iron meteorite. You measure its isotopic composition and you stick it into the equation. And you'll be famous, because you will have measured the age of the Earth.'

I said, 'Good, I will do that.'

And he said, 'It will be duck soup, Patterson.'

The piece of rock Brown supplied, a tiny lump of meteorite no bigger than a child's fist, came with an impressive pedigree. About 50,000 years earlier an enormous meteor had smashed into the Arizona desert, blowing out a hole 1,200 metres wide – the famous Meteor Crater. Although the meteor itself completely vaporised on impact, seconds before it hit, thousands of small fragments broke off and fell over the surrounding desert. Some landed three miles away from the impact crater in a ravine called Canyon Diablo, and it was one of these pieces that Patterson prepared to analyse in the spring of 1953.

Chasing a Rainbow

Before he could start, however, he had to overcome the danger of contamination. The amount of lead contained in the meteorite was so small that if he allowed even the smallest particle of modern lead into the sample, it would ruin the result. This was the biggest challenge. Lead was everywhere: in the commercial solvents he used for dissolving the mineral, in the glass from which the flasks were made, even in the air. It took him over two years to develop the clean laboratory techniques he needed for the operation. Only when he was fully satisfied that the laboratory was completely free of lead, did he begin work. He cut the meteorite in half with a diamond saw, and from the tiny black pockets of sulphide inside, dug out a small sample. He dissolved this in acid, removed the sulphide, then extracted the lead. After all the years of preparation, the whole process took no more than a day.

Since Caltech didn't yet have its own mass spectrometer, Patterson flew with the sample to Chicago to perform the analysis. Late at night, in a remote building at the Argonne National Laboratory, he attached the vial containing the tiny speck of lead to the spectrometer. He worked alone; nearly everyone else had long since headed for home. As the machine hummed, and the data spewed out onto a strip of paper, he grabbed a pen and began making rapid calculations. Subtracting the primordial from the present-day lead, he found that the amount of one isotope of lead had doubled, while the amount of the second had increased by 50 per cent. All this extra lead had come from the decay of uranium. Feeding his figures into the physicists' 'marvelous equation', he calculated the age of the Earth. It came to 4.5 billion years. 'I knew then with absolute confidence that we had it,' he wrote later.

True scientific discovery renders the brain incapable, at such moments, of shouting victoriously to the world 'Look at what *I* have done! Now *I* will reap rewards of

Clair Patterson working in his Caltech laboratory in 1957, shortly after he discovered the age of the Earth.

recognition and wealth!' Instead such discovery instinc-
tively forces the brain to thunder '*WE* did it!' in a voice
no one else can hear, within its sacred, but lonely, chapel
of scientific thought.

That September, at a conference in Williams Bay, Wisconsin,
he announced his result to the world. There was no great
euphoria at the news; there had been too many wrong estimates in the past to get too excited, and Patterson still needed
to check the figure. However in 1956, after he had confirmed
his result by other methods, the new age became widely
accepted.

As the 200-inch telescope carved its path across the heavens,
Sandage sat alone inside the prime focus cage, high above the
floor of the observatory, periodically checking the focus or
changing a plate. From a speaker in the cage came the sounds
of his favourite music, Wagnerian opera. It echoed around
the dome and out into the chill night air. Night after night,
the routine was the same.

The big advances came early. In 1954, armed with new, fast,
red-sensitive plates, Sandage went back and re-photographed
some of the galaxies Hubble had used for his distance scale.
These were the galaxies, beyond the range of cepheids, whose
distances Hubble had been forced to measure by determining
the brightness of their brightest stars. When Sandage examined the new plates, however, he was stunned. The objects
weren't stars at all, but H II regions, dense clouds of ionised
hydrogen that shone much brighter than a single star. Because
their absolute brightness was greater, the galaxies had to be
further away than Hubble had originally estimated. When
Sandage made the calculation, he trebled Hubble's original
estimate. In one leap, the age of the universe moved up to 5.5
billion years.

Further observations over the next three years revealed that even this figure was too low. In a 1957 lecture Sandage announced that the most likely value for the Hubble constant – the expansion rate of the universe – was 75 kilometres per second per megaparsec. This was seven times Hubble's original figure. The universe was ageing. But exactly by how much was hard to say. According to Sandage's margins of error, the universe could be expanding at anything between 50 and 100 kilometres per second per megaparsec. The uncertainty was enormous. Assuming that the universe had always expanded at the same rate, it would be between 10 and 20 billion years old. Translated into a probable age, based on a theoretical estimate of how quickly the expansion was slowing down, it could be as young as 7 billion years, or as old as 13 billion.

Despite its vagueness, this was the last of Sandage's great leaps forward. The difficulty in measuring accurate distances across the vast realms of empty space made it hard to pin down the figure more precisely. Eliminating this uncertainty and determining the true value of the Hubble constant was turning into a Herculean struggle. The pot of gold at the end of the rainbow remained beyond reach.

By the early 1960s the failure of Gamow's theory to account for the formation of the heavy elements had cast considerable doubt on the idea of the Big Bang. But in 1964, as the Americans raced for the Moon, two radio astronomers, Arno Penzias and Robert Wilson, stumbled across the evidence that finally convinced the world that the Big Bang had really happened.

Working at the Bell Telephone Laboratory at Holmdel just south of New York, Penzias and Wilson were testing a microwave horn antenna built to pick up satellite signals, when they detected a noise they couldn't identify. Wherever they pointed the instrument, it was always there, a constant, irritating hiss. At first they assumed the noise, like the hum from an amplifier, was coming from within their equipment, but however hard they tried, they couldn't get rid of it. Where it was coming from remained a mystery.

Then, in March 1965, after a frustrating year spent trying to eliminate the interference, Penzias happened to hear about a team of radio astronomers working just thirty miles away at Princeton, New Jersey. According to his source, they were building a radio antenna to look for the background radiation that Alpher and Herman had predicted would be left over from the Big Bang. Penzias knew nothing of Alpher and Herman's prediction, but his ears pricked up at the mention of background radiation. Suspecting that it might be the cause of the mystery noise, he phoned Princeton and described the problem to Robert Dicke, the leader of the Princeton team. As Dicke listened to Penzias's description of the rogue signal, he became more and more convinced that it was the one he and his team were looking for. The next day he and his colleagues drove over to the Holmdel antenna to inspect the evidence for themselves. By the end of the day they knew they had been beaten to the discovery. The Holmdel signal closely matched Alpher and Herman's predicted 5K radiation. Without realising it, Penzias and Wilson had found the remnants of the Big Bang.

Two months later, news of the discovery hit the front page of the *New York Times*. Without a doubt this was the biggest story in astronomy since Hubble had discovered the expanding universe, and scientists acted accordingly. When they read the details, astronomers who had previously supported Hoyle's Steady-State theory abandoned it overnight. In Rome, an ailing Lemaître learnt about the discovery just a week before he died. The picture he had imagined all those years before had come true; the fireworks of the Big Bang had left a trail of smoke after all.

*   *   *

Throughout the Sixties and Seventies, Sandage pursued his quest to pin down the Hubble constant. Night after night, he sat in his Palomar eyrie combing the heavens, checking and re-checking each step of the cosmic distance ladder. Progress was painfully slow, and although there were other rewards along the way – a breakthrough in stellar evolution, the first photograph of a quasar – the ultimate prize remained elusive.

In 1963 he teamed up with the Swiss astronomer Gustav Tammann, who became his long-term collaborator. Immaculately groomed, with a mischievous wit and easy charm, Tammann also possessed an easygoing manner that proved the perfect foil to Sandage's restless drive. The two got along well, and to this day Tammann has nothing but respect for his elder colleague. 'I think he is the greatest living astronomer,' he enthuses. 'He is the most serious person, it is essential for him; it is not a game, it is not a profession, it is his God-given duty to do the best to understand the universe. It is a vocation in him.'

Tammann's particular skill was in photometry. While Sandage probed the heavens with the Palomar telescope, Tammann pored over the resulting plates back at the Santa Barbara Street offices, searching for variable stars and measuring their magnitude.

Ironically the breakthrough, when it finally arrived, was the result of a failure. Back in the mid-Sixties they had begun to search for cepheids in the spectacular galaxy M101, popularly known as the Pinwheel, because its fragmented spiral arms resemble a Catherine-wheel throwing sparks and flames out into the night. From earlier estimates of its distance they knew it was further away than any galaxy in which cepheids had been seen before, but they still thought it would be close enough for cepheids to be visible with the 200-inch telescope. For months on end Tammann searched through Sandage's plates. 'I worked infinitely long,' he recalled. 'We worked and worked in this damn galaxy, and were, in the end, sure, to the limit of detection, there were no cepheids in that galaxy visible.' Just like Baade's RR Lyrae stars, the cepheids were definitely there; it was just that the galaxy was beyond the range of the telescope. Nevertheless, this failure helped pin down its distance. 'We had some secondary distance indicators, which were not so strong, which gave an actual estimate of the distance,' Tammann recalled, 'but we said it cannot be nearer than 23.5 million light-years, otherwise we would have seen the cepheids.'

This figure – ten times Hubble's original estimate – became their distance to the galaxy. It was also a stretch factor for everything. The more distant galaxies automatically moved even further away, the whole universe became markedly bigger, and once again the value of the Hubble constant plummeted. When, in 1975, Sandage finally announced the new figure, it was close to the bottom end of the range he had predicted fifteen years earlier. The Hubble constant, he declared, was 55, and the most probable age for the universe was approximately 15 billion years.

For twenty-five years Sandage had virtually had the field to himself. Perched in the prime focus cage of the 200-inch telescope, probing the cosmos to the strains of *Rheingold* and *Götterdämmerung*, it was easy not to notice that the outside world had moved on. The Beatles had given way to the Bee Gees, disco to punk rock. A new generation of astronomers had arrived: baby-boomers, young, brash and fiercely competitive. In the years since 1950, the number of astronomers had more than doubled. New telescopes had sprung up, many equipped with large mirrors and the latest electronic light sensors. For the first time, Sandage had serious competition; others were on the trail of the Hubble constant. His announcements, once accepted *ex cathedra*, began to come under fire.

As the number of astronomers grew, so did the number of estimates for the Hubble constant. In the forty-five years up to 1975 astronomers had made only about twenty estimates of its value. In the following five years that number more than doubled, and by 1990, there were over 100 estimates. However, with more results came more confusion. The spread of values became so wide as to be practically meaningless. Depending on who you spoke to, the Hubble constant could be as low as 41 or as high as 110. The field was thrown into a ferment of acrimony and dissent, with some observers putting the age of the universe at 6 billion years, while others put it at 16 billion. For theoreticians and others hoping to make sense of the world, the lack of a definitive value was intensely frustrating. At a time when other fundamental constants were known to within fifteen decimal places, the uncertainty in the Hubble constant, as Martin Rees, Britain's Astronomer Royal, noted a few years later, was 'embarrassing'.

There were many reasons for the discrepancies. For a start, the distance to a remote galaxy had to be built up step by step, so errors at each stage multiplied. Then, different astronomers made different assumptions – for example, on the extent to which dust in a galaxy affected the amount of light coming from a star. And last, but not least, telescopes were still not capable of resolving the fine detail necessary for making these extremely difficult measurements. The way out of this dilemma, many astronomers argued, was to build a more powerful telescope.

On the morning of 24 April 1990, as the space shuttle *Discovery* blasted off from Cape Canaveral in Florida, the hopes and dreams of the world's astronomers rested with the payload in the cargo bay: an 11-ton satellite, the size of a bus, called the Hubble Space Telescope. From its position in space, far above the distorting effects of the Earth's atmosphere, the telescope promised images of galaxies ten times sharper than any seen before. The driving force behind its design, and its primary goal, was to find the Hubble constant.

15. The Moment Time Began

There was more to finding the age of the universe, however, than just finding the Hubble constant. For sixty years astronomers had collected evidence that confirmed that the universe was expanding, swelling out from the initial Big Bang like a ripple spreading from a stone thrown into a lake. Theory also told them that this expansion should be slowing down, retarded by the gravitational pull of all the stars and galaxies, but so far, no one had been able to find the evidence for this.

This was a serious blank. To find the age of the universe, scientists needed to pin down, not just the value of the Hubble constant – which gave the present rate of expansion – but also the rate at which this expansion had slowed since the Big Bang – the deceleration. With both numbers they would be able to piece together an imaginary film of the universe's history, be able to follow its progress as the galaxies flew apart, quickly at first, then ever more slowly as gravity retarded the expansion. The universe was younger than the present rate of expansion suggested, but how much younger depended on its rate of deceleration. Only by finding both the Hubble constant and this rate of deceleration was it possible to find the true moment time began.

If finding the value of the Hubble constant was difficult, then finding the value of the deceleration ranked close to impossible. Few astronomers had ever considered the task;

those that had, soon retreated in embarrassment when their results were found to be wrong. Even Sandage had been defeated by the challenge. In the 1960s he had made an estimate based on observations of quasars (extremely bright objects at the centre of some galaxies) which tentatively suggested that the universe was slowing down so fast that it would eventually stop expanding and collapse in a Big Crunch. He had even estimated that the lifetime of the universe, from Big Bang to Big Crunch, would be 82 billion years. The idea became quite popular in the 1960s. Out of the Big Crunch would come another Big Bang, and the whole process would repeat itself. In this way, the universe could have existed for ever, bouncing in and out of existence every 82 billion years. The ancient idea of everlasting cycles was back.

It didn't last long, however. It soon emerged that quasars were not the standard candles Sandage thought they were, and he was forced to drop the idea. From then on pretty well everyone else steered clear of the subject. That is, until 1988, when Saul Perlmutter, a young physicist working at the Lawrence Berkeley Laboratory across the bay from San Francisco, decided to tackle the problem.

The problem fascinated Perlmutter; it was exactly the sort of big question that had drawn him to science in the first place. 'My original interests were always in the big fundamental questions of how the world works, I always felt like: we live in this world, someone should have given us an owner's manual.' But they hadn't. So, with the youthful ambition of a twenty-eight-year-old, Perlmutter set out to write it himself.

Shortly after his arrival at Berkeley, however, Perlmutter had realised the limitations of working as a physicist. Nearly all the big fundamental problems, the ones he longed to sink his teeth into, were tackled by vast teams: as many as three or four hundred people working together on a single experiment. 'I felt, as a graduate student, I might get a little lost in a huge particle accelerator project,' he explained, 'and it might be more fun to see if there were any other ways of addressing these really fundamental questions.' Noticing that astronomers hunted in smaller packs, he switched tracks and began teaching himself astrophysics. Instead of a particle accelerator, the universe would become his laboratory. In 1988 he joined forces with another Berkeley post-doc, Carl Pennypacker, and together they decided to attempt to measure the deceleration of the universe.

The reason so few astronomers had attempted to find the deceleration lay in the enormity of the task. To succeed, Perlmutter would have to discover how fast the universe had expanded early in its life, and compare that with its expansion rate today. He would have to look deep into its past, to a time when it was less than half its present age. He would have to measure the distance and red-shift of galaxies billions of years older than those being used to hunt the Hubble constant. It was an enormous challenge.

However, Perlmutter knew he had one advantage over those trying to pin down the Hubble constant. Unlike them, he didn't have to measure the absolute distances to these remote galaxies. All he needed were their relative distances; just knowing that a galaxy was three or four times further away than another would be enough. Combining this information with measurements of their red-shifts would enable him to find the deceleration.

But to see so far back in time required a 'standard candle' of phenomenal power – one that was sufficiently bright to be seen halfway back to the beginning of the universe. There was only one object that fitted the bill – a supernova, the bright explosion of a dying star. Although there were several different types of supernova, only one kind – Type Ia – was thought to always shine with the same luminosity.

A million times brighter than cepheids, Type Ia supernovae

can be seen a thousand times further away; they are the brightest standard candles, or 'standard bombs' – as Sandage is fond of calling them – we have. Exactly why they always explode with virtually the same absolute brightness isn't understood, but it appears to be the result of one of the strangest double acts in nature. A Type Ia supernova originates from a small dense star called a white dwarf, orbiting in a binary system with another star. Although smaller than its companion, the white dwarf is so much denser, and its gravity so much stronger, that it gradually sucks material away from its partner. It literally eats its neighbour. However, this gluttony ultimately leads the white dwarf to its own untimely end. Once it reaches a critical mass, the force of its own gravity becomes so great that it can no longer support itself and it collapses in a brilliant thermonuclear explosion – the supernova. Since every white dwarf is the same weight when it collapses, the brightness of the supernova should, in theory, always be the same. It rises to a peak about twenty days after the initial explosion, briefly shines as brightly as a whole galaxy, then, over the course of many months, slowly fades until it can be seen no more.

Perlmutter calculated that he needed to measure the redshift and luminosity of forty or fifty supernovae at different distances through space to accurately measure the deceleration. His colleagues thought he was mad. Finding one supernova was difficult enough; finding forty was virtually impossible. For a start, they are incredibly rare events; in the average galaxy you might see only three in a thousand years. Second, they only shine brightly for a few weeks; to accurately measure their brightness, he needed to catch them at their peak, in other words within weeks of the initial explosion.

Up to this time, spotting supernovae had been a matter of luck, the province of amateur astronomers, but Perlmutter planned to accomplish the near-impossible and capture them on demand. Together with Pennypacker, he devised a plan:

a two-pronged assault that would guarantee them supernovae every time they looked – or so they hoped. The aim was to photograph thousands of galaxies at once. While each galaxy might spawn only one supernova every 300 years, by looking at tens of thousands of galaxies in a single night they calculated they could catch several supernovae in one sweep.

'We built the plan for it,' recalled Perlmutter, 'and we took it to my research adviser at that time, and he didn't like it, he didn't think it was a good idea, and he didn't want us to pursue it. But one of his real strengths as a leader was he was willing to support us in doing it anyway, so he said "Why don't you guys go off and give it a try."'

The only chance of the project succeeding was to take full advantage of two recent advances in technology: the arrival of sensitive electronic cameras capable of detecting extremely faint objects, and the development of high-speed computers. While Pennypacker set to work to build a wide-field camera sensitive enough to photograph distant galaxies, Perlmutter developed the software that would scan the resulting images and spot the supernovae. The plan was to take two photographs of the same area of sky, several weeks apart. By electronically subtracting one image from the other, all the stars of equal brightness on both exposures – the regular stars – would cancel each other out, leaving visible only the newly formed supernovae.

'At first we had lots of criticisms; people didn't think we should do it, it would not work,' Perlmutter remembered. And for a long time, it looked as if they might be right. After four years' work, and observing trips to telescopes as far apart as Australia and the Canary Islands, they still hadn't found a single supernova. When they did eventually find one, in April 1992, it looked like too little, too late. In an external funding review, the prominent Harvard astronomer Robert Kirshner criticised the project for its lack of results, and questioned the viability of the whole scheme: it was too hit-or-miss, the analysis was weak, and the quality of what little data they had was poor. It was a damning and largely justified assessment that nearly sank the project for good. They only survived by the skin of their teeth.

From then on, however, things started to look up. After several trials at Kitt Peak in Arizona, they began to find batches of supernovae – two or more each run. Then in 1995, they moved to 'the best telescope for the job', the 4-metre Cerro Tololo in the Chilean Andes. 'And we basically started just racking them up,' said Perlmutter, 'just churning out groups of ten, twenty supernovae at a time.'

The secret of their success was down to teamwork, the 'group intelligence' as Perlmutter called it. For each observing run, the dozen or so people on the project split into several groups. Just after a new moon – the dark time of the month when observing conditions were best – the first team travelled to the Cerro Tololo observatory in Chile, where they photographed around 70,000 galaxies, then returned to Berkeley with the data. Type Ia supernovae take about three weeks to rise to peak brightness, so three weeks later, in the dark period just before the next new moon, they returned to Chile to photograph the same galaxies again.

Now came the race against the clock. As soon as each image came off the camera, it was sent over fast Internet lines back to base camp at Berkeley. Working through the night and long into the next day, the Berkeley team compared the new images with those taken earlier. This required number-crunching on a colossal scale, the computers scanning tens of millions of stars for the tell-tale speck of a single supernova. Each time the computer threw up a likely candidate, the team checked the observation by eye. If confirmed as a supernova, it was added to the list of targets.

The next day, target list in hand, Perlmutter and a few

colleagues jumped on a plane to Hawaii. There was no time to lose; they had to measure the red-shift of the most distant supernovae before they began to fade. On the summit of the extinct Mauna Kea volcano, the giant 10-metre Keck telescope whirred into life. From the control room at the foot of the mountain, Perlmutter and his crew guided the vast eye through the list of stars, recording their spectra and deducing their red-shifts. Three days later they returned to Berkeley exhausted, clutching the prized data in their hands.

Perlmutter had cracked it. He had shown his critics he could deliver supernovae on demand.

In the 1950s, when astronomy was still a gentlemanly pursuit, any object found by an astronomer was practically regarded as their personal property, reserved for them to probe and study at their leisure. It was considered bad form to publish a paper on someone else's star before they had had a chance to do so themselves. By the 1990s, those days were long gone. When news of Perlmutter's success began filtering through to the East Coast, an alert astronomer at Harvard began to take an interest.

In May 1994, Brian Schmidt, one of Kirshner's post-doc students, put together two vital pieces of information. 'I was sitting in an office at Harvard,' he recalled, 'and within the same week two big events happened. One, I found out that Perlmutter and company had found seven supernovae at interesting red-shifts, and at the same time I found out that the Chileans had managed to crack how to do supernovae distances accurately.' Schmidt saw straight away that although Perlmutter's team had solved how to find supernovae, they were making a crucial mistake when it came to measuring their distance.

Despite everyone's assumptions, it turned out that not all Type Ia supernovae exploded with exactly the same brightness. Some, it now emerged, shone slightly more brightly than others. What Perlmutter had missed was the recent discovery in Chile of a way to compensate for this. The previous year, a team of researchers at the Cerro Tololo observatory had found that supernovae that reached maximum brightness, then faded quickly, were not as bright as those that faded slowly. By measuring how long they took to fade, it was possible to correct for these differences in luminosity. The revelation turned Type Ia supernovae from a mediocre way of measuring distances into the sharpest tools available.

'My problem with Saul's team was they were ignoring how to use the supernovae as accurate distance indicators,' said Schmidt. 'That was half the story to me . . . I saw those two things and I said, hey, we should be doing this ourselves, because we have half the answer here.'

Working fast, Schmidt made plans for a rival supernovae search of his own. 'They showed you could really find these things, the technology was there,' he explained. 'I didn't know how they were finding them, but I knew that if they could find them, I could find them too.' By November, he had formed a loose coalition of scientists from all over the world, pulling in experienced supernovae specialists such as Kirshner together with a bunch of enthusiastic youngsters. He christened it the High-Z ('z' is the symbol astronomers use for red-shift) Supernovae Search Team.

Although latecomers to the race, Schmidt's team had the advantage of experience. Unlike Perlmutter's team – which was built up largely of physicists – they were astronomers by training. They knew the tricks of the trade, how the system worked, the right people to talk to. They understood the traps into which the unwary could fall and they knew how to avoid them. With this experience it wouldn't take them long to catch up. Sure enough, in March 1995, less than a year after starting out, they found their first supernova. The race was on.

*   *   *

The Moment Time Began

The race was on too for the Hubble constant. But only after astronauts had overcome a serious setback with the Hubble Space Telescope. Shortly after its launch, engineers had discovered a fault in its primary mirror. Instead of the clear sharp images they had expected, astronomers were faced with a blur – and an unsettling delay. It wasn't until December 1993, over three years after its launch, that a team of astronauts installed the optics needed to correct the telescope's vision. The repair bill came to around half a billion dollars.

It was with nervous anticipation that the group of astronomers chosen to use the Space Telescope to find the Hubble constant – known as the Key Project team – downloaded the first pictures from the repaired telescope. In Australia, Jeremy Mould, a co-leader of the team, was relaxing at his home when he learnt that the first images had become available. 'I can recall one Saturday afternoon after a barbecue up here, grabbing people and saying let's take a look at the data.' A small crowd gathered around the computer as he downloaded the images over the Internet. When the first picture appeared, the room erupted to whoops of delight. Instead of a faint blur, the galaxy they were looking at was lit up with millions of bright pinpricks. There was no doubt about it; the repair mission had been an outstanding success.

After years of delay, the Key Project team could finally start work. One of their main targets was M100, one of over two thousand galaxies in the giant and extremely distant group known as the Virgo cluster. The exact distance to M100 was unknown, but it was thought to be about twice as far away as M101, the Pinwheel galaxy in which Sandage had failed to find cepheids in 1975. If the Key Project team could find cepheids in M100, they could measure the distance to Virgo – a major step towards finding the true value of the Hubble constant.

At her Pasadena office, Wendy Freedman, another of the

Key Project's co-leaders, began scanning the Hubble images of M100 for cepheids. Like the supernovae search teams, Freedman and her colleagues used sophisticated computer software to pull out the cepheids from the mass of other stars. 'We were able to see the cepheids varying in brightness and measure them, and it turned out to be much easier than we had ever anticipated,' she recalled. 'That was a pretty exciting moment, because we knew that the project could be done.' From their measurements, the team estimated that M100 was 57 million light-years away, the most distant galaxy in which cepheids had ever been found. They were on their way to finding the Hubble constant.

In the middle of March 1997, Saul Perlmutter and two members of his team, hunting for the deceleration, flew into San Francisco International Airport clutching a bag of data tapes, the gleanings from three nights' supernovae hunting at the Keck Observatory in Hawaii.

It had been a nerve-racking week. In the race to find how fast the universe was slowing down, neither his team, nor the rival High-Z team, could afford to make a mistake, particularly during an observing run. The high demand for telescope time at the big telescopes like Keck, meant that on average each team could expect only two observing runs a year. They had to make the most of them. Fortunately for Perlmutter, it had been a good session: they had measured the red-shifts of nine supernovae and confirmed most of them as Type Ia. But best of all they had set a new record: around 7 billion light years away they had found the most distant supernova ever seen.

Intense professional rivalry meant that each team kept its data secret from the other, but both camps were now using similar analytical techniques, including the Chilean luminosity corrections. Perlmutter's team had the advantage of numbers – sheer quantity of supernovae – but their earlier data were weak, the result of not always being able to obtain follow-up time to check their observations. Schmidt's team, on the other hand, had fewer supernovae, but prided themselves on the accuracy of their results.

Over the summer of 1997, the two teams worked on the analysis of their spring observing runs. There were a hundred and one ways the data could give the wrong result. Dust in the host galaxy could dim the light of some supernovae more than others; so could dust in our galaxy, as well as that in intergalactic space; the clumping of matter in the universe could brighten some supernovae while dimming others; or the supernovae might not be Type Ia's at all. All these and myriad other possible errors had to be eliminated.

On 24 September, Perlmutter's team received their first hint that there was something unusual about the expansion – something deeply worrying. At their regular Wednesday meeting, Gerson Goldhaber, the eminent particle physicist who had been working on the data analysis, presented a graph to the group. The distant supernovae, he revealed, were dimmer than expected; they were further away than they should have been. From the best-fit solution to the points on the graph it looked as if the universe wasn't slowing down at all; it was speeding up.

If true, this was incredible. For seventy years nearly everyone who had considered the issue had assumed the expansion of the universe was decelerating; if it was speeding up it would be a startling discovery.

Goldhaber remembered his colleagues' reaction. 'At first there was skepticism, which is good, there should be skepticism, and slowly I persuaded the group to look at this seriously. I wanted somebody to check it and Saul assigned an undergraduate, which I was not too pleased with because I expected a senior person. But it turned out he was rather good and did manage to check it, and couldn't find any deviation from what I had put forward.'

'It was a real surprise,' said Perlmutter, who assumed they must have made a mistake. 'My main response was, let's start this experiment . . . this experiment . . . this experiment, trying to figure out: is there something wrong in this?'

Around the same time, the High-Z team were trying to come to terms with similar evidence. Back in July, new measurements from a clutch of five supernovae observed with the Hubble Space Telescope had hinted at acceleration, but at the time, Schmidt hadn't believed it. 'I was just quite frankly denying this was happening,' he admitted. 'I figured we were probably doing something wrong.' Then in November, a further batch of nine supernovae showed the same thing. 'When we went up to fourteen, fifteen objects there was just no denying it, and . . . at that point we sort of said, "Mmm, I guess the universe really is accelerating."'

By now Perlmutter was also convinced. 'Little by little, you do more and more of the calibrations, you do more and more of the cross-checks, and you check one thing after another, and finally you start believing your data.'

Both Schmidt and Perlmutter knew they were sitting on dynamite; it was the discovery of a lifetime. But both men were nervous. If they were wrong they might never live it down.

Perlmutter broke cover first. In January he made a guarded announcement at the 191st meeting of the American Astronomical Society in Washington, DC. A month later Schmidt's team made a more assertive statement at a conference at Marina Del Rey in California. *Science* magazine picked up both announcements, realised it had a major story on its hands, and issued a press release. At the beginning of March, before either group had so much as published a paper, news of the accelerating universe hit television screens across the country.

The astronomical community reacted with a mixture of astonishment and disbelief. 'I was shocked,' said Rocky Kolb, a cosmologist at the University of Chicago. 'I didn't believe it when I heard it and I still have a hard time believing it.'

'This has our minds swimming,' added astrophysicist Richard Muller. Even team members found it difficult to come to terms with the discovery. 'Amazement and horror,' said Schmidt, when asked about his own reaction. 'Amazement, because I just did not expect this result, and horror in knowing that it will likely he disbelieved by a majority of astronomers – who, like myself, are extremely skeptical of the unexpected.'

The announcement turned traditional views of the universe on their head. 'It's very weird,' said Adam Riess, the young astronomer who had processed the High-Z team's data. 'I mean it goes completely counter to our idea of what gravity does and what the universe should be doing. Gravity pulls on stuff. It would be like throwing an apple up, and having it take off and go up instead of going down. I meant it's that weird.'

Once everyone had recovered from their initial astonishment, the theoreticians began piecing together what it all meant. There were many questions that needed answering – not least, what was causing the acceleration? If Perlmutter and Schmidt were right, the universe was behaving as if it contained a hidden repulsive force, one that countered the attraction of gravity and caused the expansion to accelerate.

To some theoreticians, this wasn't as strange as it seemed. Indeed, science had been here before. Back in 1917 Einstein had proposed a remarkably similar force in order to hold the universe static – the famous cosmological constant, which he later abandoned as his 'biggest blunder'. But perhaps his 'biggest blunder' wasn't a blunder after all?

Einstein had introduced the repulsive force of the cosmological constant because he thought the universe was static. To repeat an analogy, a static universe was like a stone hovering in the air; the cosmological constant acted as a repulsive force

Accelerating Universe

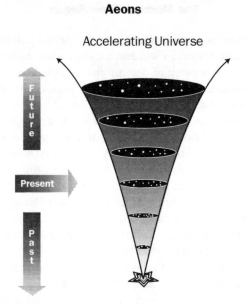

just powerful enough to balance the force of gravity – it kept the stone hanging. But suppose you made the repulsive force a bit more powerful, increased the value of the cosmological constant so it was greater than gravity; then the stone would start to rise through the air and accelerate – just like the universe. The universe was accelerating because of a force just like the one Einstein proposed. The cosmological constant was back.

But where was this force coming from?

Remarkably, the most probable explanation had come some sixty years earlier. In the 1930s, the famously taciturn English physicist Paul Dirac had shown that a vacuum, or an area of empty space, isn't really empty, but contains tiny elementary particles that pop in and out of existence too quickly to be detected directly. Later research revealed that the vacuum is teeming with a whole spectrum of these virtual particles. What looks to us like an empty void is, on the quantum level, jumping with all the activity of a hot pan of popcorn.

In 1967, Yaakov Zel'dovich, the leading Soviet astrophysicist of his generation, found that the energy contained in the vacuum would have the same effect as a cosmological constant: it would make space expand. But he was faced with an enigma. When he calculated the total amount of energy in the vacuum, it came to a staggering 10^{120} (that's a 1 followed by 120 noughts) times the total energy in the whole universe. It would blow space apart so fast that it would be expanding faster than the speed of light.

Obviously this didn't happen. But exactly why it didn't happen – why there should be such a discrepancy between observation and theory – completely confounded him and has continued to perplex particle physicists ever since. The best guess is that there is some sort of dynamic system which cancels out most of the energy of the vacuum, rendering it close to, but not actually, zero. But what this system might be, no one knows. While the vacuum energy clearly exists (its effect has been measured), and provides the best explanation for the acceleration, it is still a mystery why it lacks the enormous value quantum theory predicts.

Once the news had settled in, most scientists warmed to Perlmutter and Schmidt's discovery. 'The biggest surprise I found was how quickly people accepted it as possibly being correct,' said Perlmutter. This was largely because an accelerating universe solved more problems than it raised.

To begin with, it solved a dispute that had been simmering between the theoretical physicists and the observational astronomers for years: the problem of the missing mass. According to the theorists, the universe had to contain just enough mass to stop it expanding for ever, but not so much that it would stop expanding and collapse in on itself. It had to have what was known as the 'critical density'. However, when the observers attempted to 'weigh' the universe by adding up all the mass it appeared to contain, they found that its total mass only came to 30 per cent of this critical density. The difference – the remaining 70 per cent – was known as the missing mass.

The discrepancy had split astronomy in two. On one side were the observers who were convinced that they had added up all the mass there was; on the other were the theorists who insisted that the missing matter was out there. 'I remember when we first came up with some of these measurements,' recalled Neta Bahcall, an observer at Princeton University, 'the theoretical physicists really were quite reluctant to buy them because it was kind of an ugly universe if the mass density was not critical. So they held back, saying: "Well, yes, we see the data, but maybe there are ways out of it."'

The way out of it came with Perlmutter's discovery of the acceleration, which, remarkably, allowed both sides to be right. The missing mass, it now emerged, was not really missing, it just existed in another form. As Einstein had famously shown, mass and energy were interchangeable; they were two sides of the same coin. The reason the missing mass couldn't be seen was because it existed in the form of energy – the vacuum energy that was driving the acceleration of the universe.

And it added up. Combined with other measurements, Perlmutter and Schmidt's results enabled the amount of this repulsive energy to be calculated. To the delight of everyone, its mass equivalent turned out to be almost exactly equal to that of the missing mass. While the amount of matter in the universe came to 30 per cent of the critical density, the mass equivalent of the vacuum energy corresponded to 70 per cent. 'That is like a joke,' laughed Sandage's colleague, Gustav Tammann, scarcely able to believe the simplicity of the solution. 'You are allowed to add the two: 0.3 plus 0.7 gives 1. So at the moment it is infinitely beautiful.'

It was almost too good to be true. The observational astronomers were happy: their long-held belief that the universe contained only 30 per cent of the critical mass was finally vindicated. And the theorists were overjoyed: the universe had the critical density they had predicted after all.

On 25 May 1999, at a crowded press conference at NASA headquarters in Washington, DC, Wendy Freedman and Jeremy Mould, of the Key Project team, announced the conclusion of their search for the Hubble constant. After seven years of observations, several million measurements, and the detailed analysis of two dozen galaxies, they had finally tied the elusive figure down to within – they believed – 10 per cent of its true value. Their result put the Hubble constant at 70, and the Hubble time (the age of the universe if it had always expanded at this same rate) at 14 billion years. As one newspaper put it, 'the search for the Holy Grail of cosmology was over.'

For Allan Sandage, still working at Pasadena, the result was a bitter disappointment. For a quarter of a century he had maintained that the Hubble constant lay in the 50s, but now the balance of opinion lay against him. 'It's a disaster for me,' he told a reporter shortly after the announcement. 'It's a psychological disaster. I mean, you just can't imagine what the last two weeks have been like. But if you permit it, the world will either break your heart or turn your heart to stone.'

Fifty years earlier, when Sandage had first set out to find the age of the universe, it was thought to be just 3.6 billion years old. Since then, through his discoveries, he had added more years to its life than anyone else in history. However, the long hunt had taken its toll. Decades of staring through an eyepiece following the stars as they turned their nightly round meant he had lost some vision in his right eye, and he had stopped observing a few years earlier. But he still worked every day, still published new papers, and – when asked what he thought of his opponents' results – still defiantly insisted they were wrong. 'All of these methods have given from 90 down to 80, now they're down to 70. Our value is now 58, up from 55; so they're coming down and we're staying fairly close.'

But he was beating a lone drum. The Key Project team had strength of numbers on their side. Around the world, hundreds of astronomers had joined the hunt for the Hubble constant, and their output was prolific. In 1999, new values poured off the presses at an average rate of one a week. At first glance they still ranged as widely as ever – from a low of 44 to a high of 81 – but there were fewer of these extreme figures, which anyway tended to come from new, untested techniques. Instead, the majority of results now lay in a compact band between 63 and 73, with a median value of 68. Confirmation that the Key Project team's value of 70 was at least in the right ballpark.

With the discovery of the acceleration, and these improved estimates of the Hubble constant, astronomers now had all the information they needed to calculate the age of the universe.

In general relativity, the age of the universe depends on three numbers: the Hubble time – the time that would have elapsed since the Big Bang if the universe had always expanded at its present rate; the mass density – effectively the amount of matter in the universe – which acts to slow down the initial rate of expansion; and last, but not least, the newly discovered cosmological constant – or vacuum energy – which works to speed it up.

In Australia, Charles Lineweaver, a forty-four-year-old American physicist working at the University of New South

The Moment Time Began

Wales in Sydney, seized the opportunity to put this information together.

His task – like the history of the universe – was not straightforward. Just because the universe is accelerating now doesn't mean that it has always accelerated in the past. Ever since the Big Bang there has been a battle between gravity, which slows down the expansion, and the repulsive force associated with the vacuum energy, which speeds it up. Depending on the balance of the two forces, this can mean an accelerating universe. When the universe was young, all the matter we see today was contained in a much smaller volume of space; the universe was much denser, and consequently the gravitational attraction was much stronger. For the first several billion years of the universe's life, gravity was powerful enough to slow the expansion. But as the universe grew larger, matter thinned out, gravity's power diminished, and vacuum energy took over.

Lineweaver compares the energy inherent in the vacuum to springs. 'The springs apparently are compressed,' he explained. 'Everywhere in space they're compressed, and they would like to expand, and the more the universe expands, the more springs you have. It's a weird kind of stuff – it's defined by this weirdness. It's the structure of the vacuum and if you make the universe expand you have more of it by definition.' And so, as the universe grew, the amount of vacuum energy increased, and the expansion started to accelerate. At a point about 5 billion years ago, around the time the Earth formed, vacuum energy became the dominant force, and our universe slid gently from slowing down, to speeding up.

Surrounded by a clutter of astronomy texts and cosmology preprints, Lineweaver spent much of late 1998 and early 1999 huddled over a computer at the University of New South Wales, putting together his analysis. Into the computer went all the latest results of the new cosmology – the acceleration from Perlmutter and Schmidt's supernovae observations, half a dozen estimates of the mass density from a variety of sources, and a Hubble constant of 68, chosen largely on the strength of the Key Project team's result, but with a small nod towards Sandage's lower figure.

He also included one final, but vital, ingredient – realistic error-bars for all the results. This gave greater weight to figures that were known with high confidence, and prevented vaguely determined values over-influencing the result. It took several months ironing out bugs, correcting mistakes, updating data and repeatedly rerunning the programme before he was happy, but in April 1999 he had the most accurate figure yet for the age of the universe. According to the best estimates of modern cosmology, the moment of the Big Bang, the instant that time and the universe were created, occurred 13.4 billion years ago. For the first time in history, cosmologists had an age for the universe based on real observational data.

Given the uncertainties of the measurements, the result comes with huge error-bars (unlike Ussher, Lineweaver is a realist) which indicate it is only accurate to within plus or minus 1.6 billion years. It may lack the precision of Ussher's figure, but Lineweaver is confident that, within its margins of error, the result is accurate.

One reason for his confidence is that the figure of 13.4 billion years fits in well with the most recent estimates for the age of the oldest stars. Ever since the 1950s astrophysicists had been modelling the processes of stellar evolution, the subtle changes in brightness and colour that occur as stars burn their fuel. By the mid-1990s they were certain they knew how to calculate the age of a star. By mathematically modelling the complex nuclear reactions occurring in their cores, they could determine how long stars of different types and sizes took to burn their fuel, and from this could calculate how long they had lived. The oldest stars in our galaxy – thought to date from the time it was born – are found in the globular clusters,

The Moment Time Began

bright clumps of stars lying in a sphere around the centre of the Milky Way. Current estimates date them at 12.2 billion years. Allowing a billion or so years for the first stars to form after the Big Bang, this fits Lineweaver's date for the beginning of time perfectly.

'It almost feels, like we're taking our first baby steps as a species, as a civilization, toward actually having a model of the universe that will hold up over the next 500,000 years,' proclaimed Perlmutter, shortly after discovering the acceleration. He may be right – the present model certainly paints a coherent picture of the universe we see around us – but if there is one lesson to be drawn from history, it is that time is a harsh judge. Many theories have had their day in the limelight, only to disappear into the wings when another one appeared. Over the centuries each generation felt it had found, or was close to finding, the right answer, and it is worth remembering that Ussher was not alone in drawing the wrong conclusion about the age of the universe. Many of the greatest minds in science were equally blinkered, trapped by their own beliefs, or the prevailing assumptions of their day. Newton was every bit as religiously dogmatic as Ussher and fought to reconcile his science with the Bible; Darwin exaggerated the Earth's age to allow enough time for species to evolve; while Einstein was forced to concede that the universe was not static, but expanding – his 'greatest blunder'. Given this track record, it seems all too likely that our present picture of the universe is flawed in some way. But there are reasons for confidence. At one time the Earth appeared so large, the distance to the nearest stars so enormous, and the speed of light so fast, that none of them could possibly be measured. Yet scientists have long since measured all these with great accuracy. To modern astronomers the universe's age is no different; it is a physical reality that can be measured.

But it is more than just another measurement. In seeking to find the beginning of time, philosophers and scientists have changed our view of the world. The short timescale that was widely accepted in medieval and Renaissance times fitted the idea that God had created the world for humans. We were born with it, and would die with it when it ended. The idea that the Earth had existed without people was unthinkable. When eighteenth- and early nineteenth-century geologists discovered the unimaginable ages that had passed before humans appeared on the planet, that self-centred view was forced to change. Mankind's place in the world shrank; it became incredible that the world was created just for us. As Mark Twain wrote at the beginning of the twentieth century: 'If the Eiffel Tower were now representing the world's age, the skin of paint on the pinnacle-knob at its summit would represent man's share of that age, and anybody would perceive that that skin was what the tower was built for. I reckon they would, I dunno.' Although Twain exaggerated the briefness of humanity's existence – we would at least amount to the thickness of a 1-Euro coin – his point is still valid. We are a small and transient part of a vast and long-lived universe.

The world has not only existed much longer than was once believed, we now know that it is larger and more varied, richer and more complex, than Ussher and his contemporaries could ever have imagined. It is a universe that we have only just begun to explore, a store of untold mysteries remaining to be uncovered.

The only certainty, if history has anything to tell us, is that this latest age for the universe won't be the last. As you read this, in half the world the sky is dark, and hundreds of astronomers are working through the night, guiding their telescopes across the heavens. As night follows night, their observations grow; gradually pushing back the boundaries of our knowledge. We are charting the universe as mariners once charted the oceans, sailing out beyond the confines of the Milky Way, into the furthest reaches of the cosmos. As we probe deeper into space and lift up the veils on the past, we dream that one day, just maybe, we will know for certain the true entirety of time.

Notes and Sources

1. In the Beginning

No single book covers the full history of the search for the beginning of time, but a good starting point is Dennis R. Dean's convenient summary, 'The Age of the Earth Controversy: Beginnings to Hutton', in *Annals of Science* 38, 1981, pp. 435–56, which covers the story up to the end of the eighteenth century. Francis C. Haber's excellent *The Age of the World: Moses to Darwin*, Baltimore MD 1959, is a fuller treatment that examines, in particular, the shift from biblical to geological timescales. For a good recent treatment, see Paolo Rossi's stimulating book *The Dark Abyss of Time*, Chicago & London 1984.

The two references to Ussher's date appear in H. G. Wells, *A Short History of the World*, London 1946, p. 11, and J. B. S. Haldane, *Possible Worlds*, London 1940, pp. 18–19. The information on the cosmic cycles favoured by ancient civilisations is drawn largely from Haber, pp. 11–27, Aristotle's words come from *Meteorologica* Book 1, ch. 14, while Daniel J. Boorstin prints the first two quotations from St Augustine in his epic and entertaining history of science, *The Discoverers*, New York 1985, p. 571. For Augustine's response to the question of what God did before he created the universe, see *The City of God*, Book 11, ch. 6.

2. The Bishop and the Book

The earliest biographies of Ussher were written by two of his chaplains. Nicholas Bernard's *The Life and Death of the Most Reverend and Learned Father of our Church, Dr. James Ussher*, published in 1656, is merely an expanded version of his funeral oration, but thirty years later Richard Parr, who travelled with Ussher during the English civil war, published a fuller account in *The Life of the Most Reverend Father in God, James Ussher, late Lord Archbishop of Armagh*. A later biography

Notes and Sources

by C. A. Carr, *The Life and Times of Archbishop Ussher*, 1895, is also useful. The other main sources I have drawn from are C. R. Elrington (ed.), *The Whole Works . . . of James Ussher. With a Life of the Author, and an Account of His Writings*, 17 vols, Dublin 1829–64, and James Barr's more recent essay, 'Why the World was Created in 4004 BC: Archbishop Ussher and Biblical Chronology', in *Bulletin of the John Rylands University Library of Manchester* 67, no. 2, 1985, pp. 576–608, which explains how Ussher put together his chronology.

The lines from *As You Like It* appear in Act 4, Scene 1. Augustine's assertion that antediluvian men were giants, and his belief that they abstained from sex for over a hundred years come from *City of God*, Book 15, chs 9 and 15. Ussher's childhood progress in chronology can be found in Bernard, pp. 25–8. The first date for the age of the world appears in Theophilus, *To Autolycus*, Book 3, chs 28–9. William Hales, *A New Analysis of Chronology*, 1809, contains some 128 different dates for the Creation, but you can find some he missed in C. A. Partides, 'Renaissance Estimates of the Year of Creation', *Huntington Library Quarterly* 26, 1963, pp. 315–22. Ussher's assertion that it was possible to find the exact day of Creation is found on p. iii of his *Annals of the World*, London 1658. Carr paints a vivid picture of the difficulties of preaching in Dublin in his *Life and Times of Archbishop Ussher*, pp. 82 and 141, and Ussher's knowledge of libraries and his love of books appear on pp. 99 and 10 respectively of Parr's *Life of Ussher*. For a description of the intellectual duel with Beaumont, see Carr, p. 169. See also p. 100 for Ussher's comment that he would trust no man's eyes but his own. The American historian Anthony Grafton no doubt followed this advice in his scholarly *Joseph Scaliger, a Study in the History of Classical Scholarship*, 2 vols, Oxford 1993. Volume 2, p. 117, gives details of ancient calendar systems. The quote from John Bainbridge comes from Parr, p. 370. Josephus's reference to the eclipse in the reign of Herod can be found in *History of the Jews*, Book 17. The scribe's poignant postscript appears in John Gwynn, 'On a Syriac Manuscript Belonging to the Collection of Archbishop Ussher', *Transactions of the Royal Irish Academy* 27, Polite Literature and Antiquities, Dublin 1886, pp. 269–316. Details of how Ussher made the link with Nebuchadnezzar are given by Barr on pp. 580–1 of 'Why the World was Created in 4004 BC'. The exact hour determined by Ussher is indicated in his preface to the *Annals*, p. v: 'I encline to this opinion, that from the evening ushering in the first day of the World, to that midnight which began the first day of the Christian era, there was 4003 years, seventy dayes, and six temporarie howers.' The vivid description of Ussher witnessing the execution of Charles I comes from Parr, p. 72.

3. Doubters

Richard H. Popkin tells the fascinating story of Isaac Le Peyrère's life in *Isaac La Peyrère (1596–1676)*, Leiden 1987. The original title of La Peyrère's book was *Praeadamitae*, 1655, but within a year an English translation, *Men before Adam*, was printed in London. Some idea of the hostility the work provoked can be found in Rossi, *The Dark Abyss of Time*, p. 138, which also gives the quotation from Father Perrenin, see p. 144.

Martino Martini's travels in China and his publication of Chinese history are covered in *Historians of China and Japan*, ed. W. G. Beasley and E. G. Pulleyblank, London 1961, and in a number of works by David E. Mungello, in particular *Curious Land*, and *The Forgotten Christians of Hanzhou*, Honolulu 1994. For details of why the Jesuits had to use the Septuagint Bible, see Claudia Von Collani, 'Theology and Chronology in "Sinicae Historiae Decas Prima" ', in *Martino Martini, a Humanist and Scientist in 17th-Century China*, ed. Franco Demarchi and Riccardo Scartezzini, Trento 1996, p. 232. John Webb's unlikely explanations for why the Chinese should be the descendants of Adam and Eve are in his *An Historical Essay Endeavouring a Probability that the Language of the Empire of China is the Primitive Language*, London 1669, pp. 196–202. Jesuit attempts to check Chinese history using eclipses were brought together in P. E. Souciet (ed.), *Observations mathématiques, astronomiques, géographiques, chronologiques et physiques, tirées des anciens livres chinois; ou faites nouvellement aux Indes et à la Chine par les Pères de la Compagnie de Jésus*, Paris 1729.

4. Change and Decay

The geological discovery of time that unfolds in the following chapters has been well covered by academics and popular writers alike. For popular accounts, see Claude C. Albritton, *The Abyss of Time*, San Francisco 1980, which focuses on the individuals involved, and Ruth Moore, *The Earth We Live On*, London 1959, which although older is still a good read. In *Time's Arrow, Time's Cycle*, Penguin 1991, Stephen Jay Gould explores the contrasting ideas of linear and cyclic

Notes and Sources

time as expressed in the works of Burnet, Hutton and Lyell. For more academic works, start with Stephen Toulmin and June Goodfield, *The Discovery of Time*, London 1965, but see also Cecil J. Schneer (ed.), *Toward a History of Geology*, Cambridge MA & London 1969, and Rhoda Rappaport, *When Geologists Were Historians 1665–1750*, New York 1997. Two works that concentrate on the history of British geology are Gordon L. Davies, *The Earth in Decay: a History of British Geomorphology 1578–1878*, New York 1969, and Roy Porter, *The Making of Geology: Earth Science in Britain*, Cambridge 1977.

Turning to the details of this chapter, Descartes's struggle to find a rational explanation for how the universe was created is well told in Jack R. Vrooman, *René Descartes*, New York 1970, and in William R. Shea, *The Magic of Numbers and Motion: The Scientific Career of René Descartes*, Canton MA 1991. A translation of Descartes's *Principles of Philosophy* can be found in *The Philosophical Writings of Descartes* 1, trans. John Cottingham *et al.*, Cambridge 1985. Descartes's influence on the scholars of Cambridge is illuminated in Richard Westfall's masterly *Never at Rest, a Biography of Isaac Newton*, Cambridge 1980, p. 90.

My chief sources for information on Burnet were the ever-useful *Dictionary of Scientific Biography*, 16 vols, ed. Charles Coulston Gillespie, New York 1970–80, and Marjorie Hope Nicolson, *Mountain Gloom and Mountain Glory: The Development of the Aesthetics of the Infinite*, Ithaca NY 1959. The reference for St Peter's statement is 2 Peter 3: 8. Newton's letters to Burnet appear in Isaac Newton, *Correspondence* 2, ed. H. W. Turnbull, Cambridge 1960, pp. 329–34. Evelyn's reaction to Burnet's theory is quoted by Nicolson in *Mountain Gloom*, p. 189, while the reaction of Herbert Croft appears in his book, *Some Animadversions upon a Book Intituled the Theory of the Earth*, London 1685, p. 178. The verse satirising Burnet can be found in William King, *The Original Works of William King*, 3 vols, London 1776, vol. 1, p. ix.

For the two British naturalists who glimpsed, but failed to appreciate, the depth of the abyss, see Charles E. Raven, *John Ray, Naturalist*, Cambridge 1942; Frank Emery, *Edward Lhuyd FRS 1660–1709*, Caerdydd 1971; and R. T. Gunther, *Early Science in Oxford*, vol. XIV, *Life and Letters of Edward Lhwyd*, Oxford 1945. Lhuyd's description of the boulders at Nant-Phrancon near Llanberis appears on p. 158 of this last book. Ray recounted his visit to Bruges in his *Observations, Topographical, Moral & Physiologicall; Made in a Journey through Part* of the Low-Countries, Germany, Italy, and France: etc., London 1673, p. 7. Several of the quotations in this section come from *Further Correspondence of John Ray*, ed. R. W. T. Gunther, London 1928 (see pp. 267 and 259). Finally, Ray's observation of the orderly shell beds that cast doubt on the duration of the Flood can be found in his *Three Physico-Theological Discourses*, London 1693, p. 132.

5. God's Force

John Conduitt's account of Newton and the apple is one of the earliest versions of this story, and can be found in Westfall, *Never at Rest*, p. 154. Those keen to learn more about Woodward will find Roy Porter's 'John Woodward: "A Droll Sort of Philosopher" ', *Geological Magazine* 116 (Sept. 1979), pp. 335–417, a good starting place. A shorter biography is provided by V. A. Eyles in 'John Woodward, FRS . . . A Bio-bibliographical Account of His Life and Work', *Journal of the Society for the Bibliography of Natural History* 5 (1971), pp. 399–427. Woodward published his theory in *An Essay toward a Natural History of the Earth: and terrestrial bodies, especially minerals: as also of the sea, rivers and springs. With an account of the Universal Deluge and of the effects it had upon the Earth*, London 1695. The extraordinary story of Johann Scheuchzer and his fossil 'man' is told by Melvin E. Jahn in 'Some Notes on Dr Scheuchzer and on *Homo diluvii testis*', *Toward a History of Geology*, pp. 193–212. The reports of 'giants' found in America and France come from Raymond P. Stearns, *Science in the British Colonies of America*, Urbana 1970, p. 412, and Andrew Dickson White, *A History of the Warfare of Science with Theology in Christendom*, New York 1898, ch. 5, part 3.

For the life of William Whiston see James E. Force, *William Whiston, Honest Newtonian*, Cambridge 1985, and William Whiston, *Memoirs of the Life and Writings of Mr William Whiston*, 2 vols, London 1753. Whiston links the Flood to the comet of 1680 in *The Cause of the Deluge Demonstrated: Wherin it is Proved that the Famous Comet of 1680 came by the Earth at the Deluge and was the Occasion of it. Being an Appendix to the Second Edition of the New Theory of the Earth*, London 1708, pp. 3–6.

A full account of Newton's attempts at chronology can be found in Frank Manuel, *Isaac Newton, Historian*, Cambridge 1963. For a concise report see Derek Gjertsen, *The Newton Handbook*, London 1986, pp. 112–15, or Boorstin's colourful description in *The Discoverers*, pp.

Notes and Sources

600–3. When Voltaire read Newton's *Chronology of Ancient Kingdoms Amended*, London 1728, he gave an open verdict. See *Letters on England*, Penguin 1980, pp. 86–91. However Westfall, *Never at Rest*, pp. 804–15, is damning: 'A work of colossal tedium . . . It is read today only by the tiniest remnant who for their sins must pass through its purgatory.' Richard Bentley's verse linking Newton and Woodward is printed in Porter, 'John Woodward', *Geological Magazine* 116, p. 339. Edmond Halley's paper, 'A Short Account of the Cause of the Saltness of the Ocean', can be found in *Philosophical Transactions of the Royal Society* 29, 1715, pp. 296–300.

6. The Heat Within

The most recent, and best, biography of Buffon is Jacques Roger, *Buffon: a Life in Natural History*, trans. Sarah Bonnefoi, Ithaca and London 1997, though two others – Otis E. Fellows and Stephen F. Milliken, *Buffon*, New York 1972, and an earlier work by Léon Bertin *et al.*, *Buffon*, Paris 1952 – are also useful. Marie-Jean Hérault de Séchelles, a visitor to Buffon's estate, wrote a lively first-hand account of his encounter with Buffon in *Voyage à Montbard, fait en 1785*, Paris 1890. An early translation of Buffon's *Histoire naturelle* into English, *Natural History*, was published in London in 1792, and while no edition of *Les Époques de la nature* exists in English, a recent critical edition with an introduction by Jacques Roger can be found in *Mémoires du Muséum National d'Histoire Naturelle* série C, Sciences de la Terre, vol. 10, Paris 1962.

For the account of a comet retaining heat that inspired Buffon's theory, see Newton, *The Mathematical Principles of Natural Philosophy*, vol. 2, London 1729, p. 361. Jean Richer's observations of his slow-running pendulum are detailed in his narrative *Observations astronomiques et physiques faites en l'isle de Caïenne*, Paris 1679. Voltaire's comparison of French and English philosophy can be found in *Letters on England*, p. 68. The translation of the 1751 letter from the Sorbonne comes from Roger, *Buffon*, p. 188. For the charming story of *Telliamed* see Benoît de Maillet, *Telliamed*, trans. Albert V. Carozzi, Urbana IL 1968. This version includes detailed and fascinating footnotes and a short biography. Information on Linneus and the land rising in Sweden came from Tore Frängsmyr, 'Linneus as a Geologist', in *Linneus, the Man and his Work*, ed. Tore Frängsmyr, Berkeley & London 1983, p. 127.

7. Layers of Time

A short version of Patrick Brydone's popular travel book is still available in Italian. The original English version is *A Tour through Sicily and Malta*, 2 vols, London 1773. Dr Johnson's response is found in James Boswell, *The Life of Samuel Johnson*, 6 vols, Oxford 1964, vol. 2, pp. 467–8, and the Bishop of Llandaff's in Richard Watson, *An Apology for Christianity*, Dublin 1777, p. 174. For the impact the book had on Sicily, see John Ingamells, *A Dictionary of British and Irish Travellers in Italy, 1701–1800*, New Haven & London 1997. Josiah Wedgwood's letter is printed in Humphrey Jennings, *Pandaemonium*, Picador Books, 1985, p. 63. The original source for most information on Hutton, particularly his later life, is John Playfair's *Biographical Account of James Hutton*, Edinburgh 1797. However, Hutton's character remained elusive until I read J. Jones, H. S. Torrens and E. Robinson, 'The Correspondence of Hutton and Watt/Clerk-Maxwell', *Annals of Science* vol. 51, 1994, pp. 637–53; vol. 52, 1995, pp. 357–82. The bawdy tone of Hutton's letters is all the more surprising when one compares them with the sober prose of his scientific works. In chronological order, these begin with *The 1785 Abstract of James Hutton's Theory of the Earth*, introduction by G. Y. Craig, Edinburgh 1997; 'Theory of the Earth; or an Investigation of the Laws Observable in Composition, Dissolution, and Restoration of Land upon the Globe', *Transactions of the Royal Society of Edinburgh* 1, 1788; *Theory of the Earth with Proofs and Illustrations*, 2 vols, Edinburgh 1795. His famous quote 'we find no vestige of a beginning' appears in *Transactions*, vol. 1, p. 96. The quotation from Priestley comes from Roy Porter, *Enlightenment Britain and the Creation of the Modern World*, London 2000, p. 415.

8. A Fossil Clock

Charles Lyell's attempt to construct a fossil chronometer is well documented in two articles by Martin J. S. Rudwick: 'Lyell on Etna, and the Antiquity of the Earth', in Cecil J. Schneer (ed.), *Toward a History of Geology*, Cambridge MA and London 1969, pp. 288–304, and 'Charles Lyell's Dream of a Statistical Palaeontology', *Palaeontology* 21, 1978, pp. 225–44. However, Rudwick best conveys the broader context of Lyell's work and his disagreement with William Buckland in his excellent *The Meaning of Fossils*, London & New York 1972, pp. 164–80. For details of Lyell's journey through France and Italy I have drawn mainly from Leonard G. Wilson, *Charles Lyell, the Years*

Notes and Sources

to 1841: The Revolution in Geology, New Haven & London 1972, and Lyell's correspondence in Katherine M. Lyell (ed.), *Life, Letters and Journals of Sir Charles Lyell, Bart.*, 2 vols, London 1881. Conveniently, Lyell's major work *Principles of Geology*, 3 vols, Chicago 1990–3, has recently been republished with a useful introduction by Rudwick.

A lively account of the eccentric life of William Buckland is given by his daughter, Elizabeth Gordon, in *The Life and Correspondence of William Buckland . . .* , London 1894. For a more focused view see Nicolaas A. Rupke, *The Great Chain of History: William Buckland and the English School of Geology (1814–1849)*, London & New York 1983. Sir James Hall's attempts to trigger waves by explosives can be found in James Hall, 'On the Revolutions of the Earth's Surface', *Transactions of the Royal Society of Edinburgh* 7, 1812, pp. 139–210. George Poullet Scrope wrote up his study of the *puys* in the Auvergne in *The Geology and Extinct Volcanoes of Central France*, London 1858. His echoing refrain ' – Time! – Time! – Time!' is found on p. 209. Although extremely cautious, Lyell produced two ages for Etna. In a letter to his sister dated 19 Oct. 1830 he gave the figure of 100,000 years (see Paul Tasch, 'A Quantitative Estimate of Geological Time by Lyell', *Isis* 66, 1975, p. 406), while the second figure of 70,000 years comes from a lecture he gave at King's College on 5 June 1832 (see Wilson, *Charles Lyell, the Years to 1841*, p. 359). Lyell, *Principles of Geology*, vol. 3, pp. 97–101, gives details of his method, but no date. Switching back to Buckland, the 'clear as mud' verse describing his theories comes from F. J. North, 'Paviland Cave, the "Red Lady", the deluge and William Buckland', *Annals of Science* 5, 1942, pp. 91–128. For the Church's reaction to the discovery of the earth's old age I have drawn heavily on Haber, *The Age of the World*, pp. 201–12.

9. Burnt Fingers

Darwin's misguided attempt to calculate the age of the Weald, and the backlash it provoked, are engagingly told in Joe D. Burchfield, 'Darwin and the Dilemma of Geological Time', *Isis* 65, 1974, pp. 301–21. For an excellent and absorbing biography of Darwin, I thoroughly recommend Adrian Desmond and James Moore's *Darwin*, published in paperback by Penguin in 1992. Darwin's letters have appeared in a number of collections over the years, but the definitive (though not yet complete) series is *Correspondence of Charles Darwin*, ed. Burkhardt and Smith, Cambridge 1985–.

Herschel's letter to Lyell is reprinted in W. F. Cannon, 'The Impact of Uniformitarianism: Two Letters from John Herschel to Charles Lyell, 1836–37', *Proceedings of the American Philosophical Society* 105, 1961, pp. 301–14. Darwin's mention of the accepted 6,000-year timescale for the world appears in a letter to his sister, Caroline Darwin, dated 27 Feb. 1837. The public reaction to the dinosaurs of Crystal Palace is captured in Adrian Desmond, 'Central Park's Fragile Dinosaurs', *Natural History* 83, pp. 64–71. Darwin's first Weald calculation dates from 14 Sept. 1856 (Darwin Archive manuscript: 205(9)3:316); his second from March 1857 (manuscript 205(9)3:346); and his third from 23 Dec. 1858 (manuscript: 205(9)3:341). His argument for the longevity of geological time appears in *The Origin of Species*, Penguin 1985, pp. 293–7. Modern estimates put the age of the Weald at around 135 million years. John Phillips' estimate for the age of the sedimentary strata can be found in John Phillips, *Life on Earth, Its Origin and Succession*, Cambridge 1860, p. 126. The data on which he based his result came from Robert Everest, 'Some Observations on the Quantity of Earthy Matter brought down by the Ganges River', *Journal of the Asiatic Society of Bengal* 1, 1832, pp. 238–42. Darwin's letter to Lyell warning him not to burn his fingers is dated 25 Nov. 1860. Leonard Horner's plea for publishers to stop printing Ussher's dates in the Bible can be found in his 'Anniversary Address of the President, at the Annual General Meeting 15th February 1861', *Proceedings of the Geological Society* 17, 1861, pp. lxviii–lxx. Darwin's surprise that Ussher's date was not part of the Bible comes from a letter to Horner dated 20 Mar. 1861.

10. Doomsday Postponed

Joe Burchfield has published an exhaustive account of the long-running disagreement between Kelvin, Darwin and their respective supporters in *Lord Kelvin and the Age of the Earth*, Chicago & London 1990. The three major biographies I have consulted for Kelvin's life are: Silvanus P. Thompson, *Life of William Thomson, Baron Kelvin of Largs*, London 1910; Agnes Gardner King, *Kelvin the Man*, London 1925; and the more recent work by Crosbie Smith and M. Norton Wise, *Energy and Empire: A Biographical Study of Lord Kelvin*, Cambridge 1989. Kelvin's views on Darwin's supporters are expressed in David Wilson, *William Thomson, Lord Kelvin. His way of teaching natural philosophy*, Glasgow 1910. Kelvin's first argument to limit the age of the world can be found in 'On the Age of the Sun's Heat', *Macmillan's Magazine* 5, March 1862, pp. 388–9. His second line of attack, his paper 'On the Secular Cooling of the Earth', appears in *Philosophical Magazine* series 4, vol. 25, no. 165, Jan. 1863, pp. 1–14. The story of Fourier's astonishment at the age of the cooling earth comes from John Gribbin, *The Birth of Time*, London 1999, pp. 13–14. Darwin expresses his hopes for the future of humanity in a letter to Lyell dated 27 Apr. 1860, and his horror at the prospect of a freezing world in another to Hooker dated 9 Feb. 1865. Kelvin's third approach, linking the age of the earth to the size of the bulge at the Equator, can be found in 'On Geological Time', *Popular Lectures*, 3 vols, 1891–4, vol. 2, pp. 10–64. Nowadays it is thought that tidal retardation lengthens the day by only 2 seconds every 100,000 years (Don L. Eicher, *Geologic Time*, Englewood Cliffs NJ 1968), and that due to the molten nature of the outer core, the bulge was not fixed when the earth formed. Kelvin's reaction to the removal of Ussher's dates in the Revised Version of the Bible appears on p. 30 of King, *Kelvin the Man*. The years in which Oxford and Cambridge University presses removed Ussher's marginal notes are given in J. F. Kirkaldy, *Geological Time*, Edinburgh 1971, p. 5. Kelvin's lecture at the Victoria Institute, where he attempted to limit the age of the world to just 20 million years, appears in 'The Age of the Earth as an Abode Fitted for Life', *Science* 9, 12 May 1899, pp. 665–74; 19 May 1899, pp. 704–11.

The reason so many estimates based on sedimentation agreed with Kelvin's figure of 100 million years appears to be because that was the result geologists expected to find. The remarkable agreement between Kelvin's 100 million years and Joly's result, however, appears to have been sheer coincidence. We now know that as fast as rivers add sodium to the sea, it is either extracted into the air or falls to the sea floor in sedimentation; it has reached equilibrium. What Joly actually calculated was the average time sodium remains in the ocean before being removed. In this, his figure of 90 million years wasn't far off the mark – the current estimate is 68 million years (G. Brent Dalrymple, *The Age of the Earth*, Stanford CA 1991, p. 58).

The imaginative if somewhat dangerous early uses of radium are described in Lawrence Badash, 'The Origins of Big Science at McGill', in *Rutherford and Physics at the Turn of the Century*, ed. Mario Bunge and William R. Shea, New York 1979, p. 28, and Richard F. Mould, *A History of X-rays and Radium*, London 1980, p. 9. Rutherford's life is documented in two useful biographies: A. S. Eve, *Rutherford*, Cambridge 1939, and David Wilson, *Rutherford, Simple Genius*, London 1983. His discovery, with Soddy, of 'transmutation' appears in Muriel Howarth, *Pioneer Research on the Atom: The Life Story of Frederick Soddy*, London 1958, and Thaddeus J. Trenn, *The Self-Splitting Atom. The History of the Rutherford–Soddy Collaboration*, London 1977. Alfred Romer reprinted their groundbreaking paper of 1903 in *The Discovery of Radioactivity and Transmutation*, New York 1964, pp. 151–66. For Rutherford's account of his lecture at the Royal Institution, see Eve, *Rutherford*, p. 107.

11. Rock of Ages

The exact chain of events that triggered Rutherford's realisation that radioactivity could be used to date rocks is unclear, but William Ramsay and Frederick Soddy's discovery that radium gave off helium was crucial. They describe it in two papers, 'Experiments in Radioactivity, and the Production of Helium from Radium', *Proceedings of the Royal Society of London* 72, 1903–4, pp. 204–7, and 'Further Experiments on the Production of Helium from Radium', *Proceedings of the Royal Society of London* 73, 1904, pp. 346–58. Soddy's vivid description of their experiment appears in Morris W. Travers, *A Life of Sir William Ramsay*, London 1956, p. 214. I've drawn the portrait of the St Louis World's Fair from two sources: Robert W. Rydell, *All the World's a Fair*, Chicago & London 1984, pp. 155–83, and the official record of the exhibition, *History of the Louisiana Purchase Exposition – St Louis World's Fair of 1904*, St Louis 1905. Linda Merricks quotes Soddy's apocalyptic recollection of the exhibition in *The World Made New: Frederick Soddy, Science, Politics, and Environment*, Oxford 1996, p. 167. Rutherford's paper delivered at the conference appears in 'Present Problems in Radioactivity', *Popular Science Monthly* 67, 1905, pp. 5–34, and *Congress of Arts and Science: Universal Exposition, St Louis 1904. Division C. Physical Science*, Boston and New York 1906, vol. 4, pp. 157–86.

The next stage of Rutherford's work – his development, with Boltwood, of the lead method of radioactive dating – is summarised in Lawrence Badash, 'Rutherford, Boltwood, and the Age of the Earth: the Origin of Radioactive Dating Techniques', *Proceedings of the American Philosophical Society* 112, 1968, pp. 157–69. Their step-by-step progress can be followed in *Rutherford and Boltwood: Letters on Radio-*

Notes and Sources

activity, ed. Lawrence Badash, New Haven and London 1969. John Joly's work with Rutherford on zircon haloes is described by Arthur Holmes in *The Age of the Earth*, London 1927, pp. 67–70. Joly's first-hand account of the Easter Rebellion can be found in John Joly, *Reminiscences and Anticipations*, London 1920, p. 245.

12. Star Gazing

There are several excellent popular books on the history of twentieth-century astronomy. A good starting point is Timothy Ferris, *The Red Limit*, New York 1977, which describes 'the search for the edge of the universe', or Barry Parker, *Creation: The Story of the Origin and Evolution of the Universe*, New York and London 1988, which covers similar ground in a slightly more popular style. In *The Birth of Time*, London 1999, the astronomer and science writer John Gribbin provides a personal account of his attempt to determine the Hubble constant, and at the same time puts the search into its historical perspective.

Turning to the details of this chapter, for a brief history of philosophers' attempts to discover whether light travelled instantaneously or over time, see I. B. Cohen's fascinating article, 'Roemer and the First Determination of the Velocity of Light (1676)', *Isis* 31, 1940, pp. 327–79. Thomas Campbell's astonishment at Herschel's announcement of the enormity of space comes from Haber, *The Age of the World*, p. 228. Bessie Zaban Jones and Lyle Gifford Boyd provide a useful description of Henrietta Leavitt's work in *The Harvard College Observatory, The First Four Directorships, 1839–1919*, Harvard University Press 1971, while the work of Harlow Shapley is succinctly summarised by Owen Gingerich and Barbara Welther in 'Harlow Shapley and the Cepheids', *Sky and Telescope* 70, 1985, pp. 540–2. Shapley's own account of his life and work appear in his book *Through Rugged Ways to the Stars*, New York 1969. The definitive biography of Edwin Hubble is Gale E. Christianson's compelling *Edwin Hubble, Mariner of the Nebulae*, Bristol and Philadelphia 1995. Hubble's letters and selected papers, along with an unpublished biography written by his wife Grace, are conveniently available on a series of microfilms from the Huntington Library.

13. Expanding Horizons

How theory and observation came together in the discovery of the expanding universe is told in great detail in Robert W. Smith, *The Expanding Universe. Astronomy's 'Great Debate' 1900–1931*, Cambridge 1982. Other useful accounts are Alexander S. Sharov and Igor D. Novikov, *Edwin Hubble, the Discoverer of the Big Bang Universe*, Cambridge 1993, and Helge Kragh's comparative study of the development of the Big Bang and Steady State theories, *Cosmology and Controversy*, Princeton NJ 1996. For the life of Alexander Friedmann see E. A. Tropp, V. Ya. Frenkel and A. D. Chermin, *Alexander A. Friedmann: The Man Who Made the Universe Expand*, Cambridge 1994. For the life and work of Georges Lemaître see André Deprit, 'Monsignor Georges Lemaître', in *The Big Bang and Georges Lemaître*, ed. A. Berger, Dordrecht, Boston & Lancaster 1984, and O. Godart and M. Heller, *Cosmology of Lemaître*, Tucson 1985. Lemaître's two truths quotation comes from Duncan Aikman, 'Lemaître Follows Two Paths to Truth', *The New York Times Magazine*, 19 Feb. 1933, pp. 3, 18, while his encounter with Einstein at the 1927 Solvay Congress can be found in Helge Kragh, 'The Beginning of the World: Georges Lemaître and the Expanding Universe', *Centaurus* 32, 1987, pp. 114–39.

Finally, Hubble and Humason's 1931 estimate of the rate of expansion appears in Edwin Hubble and Milton L. Humason, 'The Velocity–Distance Relation Among Extra-Galactic Nebulae', *Astrophysical Journal* 74, 1931, pp. 43–80.

14. Chasing a Rainbow

I am grateful to Allan Sandage not only for help on the details of his own research but also for pointing me towards further sources of information. His own dogged quest for the Hubble constant is brilliantly conveyed in *Lonely Hearts of the Cosmos*, London 1991, Dennis Overbye's gripping account of astronomy in the mid to late twentieth century. An interview with Sandage can be found in Alan Lightman and Roberta Brawer, *Origins: The Lives and Worlds of Modern Cosmologists*, Cambridge, Mass., 1990. A useful summary of the work of the Pasadena astronomers is contained in Allan Sandage, 'The First 50 Years at Palomar: 1949–1999', *Annual Review of Astronomy and Astrophysics* 37, 1999, pp. 445–86, while Baade's work on the distance scale is revealed in Donald E. Osterbrock's illuminating article, 'Walter Baade, Observational Astrophysicist', in *Journal for the History of Astronomy* 27, 1996, pp. 301–48. Baade's announcement of his new age for the universe appears in *Transactions of the International Astronomical Union* 8, 1952, pp. 397–9.

George Gamow's entertaining autobiography *My World Line*, New

Notes and Sources

York 1970, tells the story of his early life and briefly touches on his work with Ralph Alpher and Robert Herman. They in turn provide a vivid picture of his extrovert personality in 'Memories of Gamow', *George Gamow Memorial Volume*, ed. Frederick Reines, London 1972. A useful summary of the scientific work of the three collaborators can be found in Alpher and Herman, 'Reflections on Early Work on "Big Bang" Cosmology', *Physics Today* 41, 1988, pp. 24–34. Gamow's early ideas are contained in his short paper 'Expanding Universe and the Origin of Elements', *Physical Review* 70, 1946, pp. 572–3, while the famous joint paper with Alpher (commonly known as the alpha, beta, gamma paper because Gamow mischievously added the name of the physicist Hans Bethe without his knowledge) is R. A. Alpher, H. Bethe and G. Gamow, 'The Origin of the Chemical Elements', *Physical Review* 73, 1948, pp. 803–4.

The worldwide reaction to Lemaître's ideas, particularly that of the Soviet Union and the Vatican, can be found in Kragh, *Cosmology and Controversy*. The horrific events of the 1937 purge of physicists are revealed all too clearly in Gennady E. Gorelik and Victor Ya. Frenkel, *Matvei Petrovich Bronstein and Soviet Theoretical Physics in the Thirties*, trans. Valentina M. Levina, Basel, Boston & Berlin 1994, and Paul R. Josephson, *Physics and Politics in Revolutionary Russia*, Berkeley, Los Angeles & Oxford 1991.

For Clair Patterson's discovery of the age of the Earth I have drawn on several sources. I would especially like to thank Patterson's widow, Lorna Patterson, for generously taking the time to talk to me, and for sending me the relevant sections of Patterson's unfinished autobiographical notes. Shirley Cohen's 1995 interview with Patterson for the Caltech Archives Oral History Project is another helpful source. A condensed version appears in 'Duck Soup and Lead', *Engineering & Science* (Caltech Alumni Magazine), vol. LX, no. 1, 1997, pp. 21–31. Patterson's reaction to his discovery can be found in Clair C. Patterson, 'Historical Changes in Integrity and Worth of Scientific Knowledge', *Geochim. Cosmochim. Acta* 58, 1994, pp. 3141–3. For those seeking more technical information, Patterson announced his result in 'The Isotopic Composition of Meteoritic, Basaltic and Oceanic Leads, and the Age of the Earth', *Proceedings of the Conference on Nuclear Processes in Geologic Settings, Williams Bay, Wisconsin*, 21–23 Sept. 1953, pp. 36–40.

Sandage's estimate in the late 1950s that the universe was between

7 and 13 billion years old appears in Allan Sandage, 'Current Problems in the Extragalactic Distance Scale', *Astrophysical Journal* 127, 1958, pp. 513–25, and the *New York Times*, 2 Nov. 1958, section IV, p. 9.

A lively account of the discovery of the cosmic background radiation can be found in Ferris, *The Red Limit*, pp. 131–51, while Arno A. Penzias, 'Cosmology and Microwave Astronomy', in Reines (ed.), *George Gamow Memorial Volume*, gives a first-hand account.

15. The Moment Time Began

For this chapter I am indebted to the many astronomers who generously spared the time to talk about their research, in particular Saul Perlmutter at the Lawrence Berkeley National Laboratory, Brian Schmidt at the Mount Stromlo and Siding Spring Observatories, and Charles Lineweaver at the University of New South Wales. The story of the discovery of the accelerating universe appears in a number of magazine and newspaper articles dating from January 1998. James Glanz's reports in *Science*, notably 'Astronomers See a Cosmic Antigravity Force at Work', *Science* 279, 27 Feb. 1998, pp. 1298–9, appear to have brought the story to the attention of the national press. Good feature-length articles are John Noble Wilford, 'In the Light of Dying Stars, Astronomers See Intimations of Cosmic Immortality', *New York Times*, 21 Apr. 1998, p. F1, and Kathy Sawyer, 'Cosmic Driving Force?; Scientists' Work on "Dark Energy" Mystery Could Yield a New View of the Universe', *Washington Post*, 19 Feb. 2000, p. A01. Among the compelling portraits of scientists at work contained in Ted Anton's *Bold Science*, New York 2000, is a whole chapter devoted to Saul Perlmutter and the Supernova Cosmology Project.

The 'other measurements' mentioned on p. 270 that allowed the value of the repulsive energy to be calculated were the latest measurements of the cosmic microwave background. These also played an important part in Lineweaver's calculation of the age of the universe.

For Sandage's reaction to the Key Project team's announcement of their value of the Hubble constant, see Richard Panek, 'The Loneliness of the Long-Distance Cosmologist', *New York Times Magazine*, 25 July 1999, pp. 22–5. Anyone interested in following the history of the Hubble constant can see how its value has changed over the years in a list of over 300 estimates compiled by John Huchra at: http://cfa-www.harvard.edu/~huchra/hubble.plot.dat.

Charles Lineweaver announced his result in 'A Younger Age for the

Universe', *Science* 284, 28 May 1999, pp. 1503–7. Two useful articles which explain current thinking on the nature of the universe are Lawrence M. Krauss, 'Cosmological Antigravity', *Scientific American*, Jan. 1999, pp. 35–41, and Neta A. Bahcall *et al.*, 'The Cosmic Triangle: Revealing the State of the Universe', *Science* 284, 28 May 1999, pp. 1481–8.

Finally, the quotation from Mark Twain comes from 'Was the World Made for Man?', *Letters from the Earth*, 1962, pp. 211–16. Our share of history isn't quite as small as Mark Twain made out. *Homo habilis*, the earliest human species, is believed to have emerged some 2 million years ago, while our own species, *Homo sapiens*, first appeared around 100,000 years ago. Using the Eiffel Tower to represent the 13.4 billion years since the universe began, the first humans arrived 4.5 centimetres from the top, while *Homo sapiens* has been around for the last 2.2 millimetres.

Select Bibliography

Albritton, Claude C. *The Abyss of Time* (San Francisco: Freeman, Cooper & Company, 1980)

Anton, Ted. *Bold Science: Seven Scientists who are Changing our World* (New York: W. H. Freeman and Company, 2000)

Barr, James. 'Why the World was Created in 4004 BC: Archbishop Ussher and Biblical Chronology', *Bulletin of the John Rylands University Library of Manchester* 67, no. 2, 1985, pp. 576–608

Brydone, Patrick. *A Tour through Sicily and Malta*, 2 vols (London 1773)

Burchfield, Joe D. 'Darwin and the Dilemma of Geological Time', *Isis* 65, 1974, pp. 301–21

——*Lord Kelvin and the Age of the Earth* (Chicago & London: The University of Chicago Press, 1990)

Burnet, Thomas. *Sacred Theory of the Earth* (London: R. Norton, 1691)

——*The Sacred Theory of the Earth* (Carbondale: Southern Illinois University Press, 1965)

Christianson, Gale E. *Edwin Hubble, Mariner of the Nebulae* (Bristol & Philadelphia: Institute of Physics Publishing, 1995)

de Maillet, Benoît. *Telliamed* trans. Albert V. Carozzi (Urbana, Chicago & London: University of Illinois Press, 1968)

Darwin, Charles. *The Origin of Species* (London: Penguin, 1985)

Select Bibliography

Dean, Dennis R. 'The Age of the Earth Controversy: Beginnings to Hutton', *Annals of Science* 38, 1981, pp. 435–56

Desmond, Adrian, and Moore, James. *Darwin* (London: Penguin, 1992)

Eicher, Don L. *Geologic Time* (Englewood Cliffs, NJ: Prentice-Hall, Inc., 1986)

Eve, A. S. *Rutherford* (Cambridge: Cambridge University Press, 1939)

Ferris, Timothy. *The Red Limit* (New York: William Morrow and Company, Inc., 1977)

Gordon, Elizabeth O. *The Life and Correspondence of William Buckland* (London: John Murray, 1894)

Gould, Stephen Jay. *Time's Arrow, Time's Cycle* (London: Penguin, 1991)

Gribbin, John. *The Birth of Time: How We Measured The Age of the Universe* (London: Weidenfeld & Nicolson, 1999)

Haber, Francis C. *The Age of the World: Moses to Darwin* (Baltimore: The Johns Hopkins Press, 1959)

Kragh, Helge. *Cosmology and Controversy. The Historical Development of Two Theories of the Universe* (Princeton NJ: Princeton University Press, 1996)

Lyell, Charles. *Principles of Geology*, with a new introduction by Martin J. S. Rudwick, 3 vols, facsimile reprint of the 1st edition: London: John Murray, 1830–3 (Chicago & London: University of Chicago Press, 1990–3)

——*Principles of Geology*, edited and with an introduction by James A. Secord (London: Penguin 1997)

Moore, Ruth. *The Earth We Live On* (London: Jonathan Cape, 1959)

Overbye, Dennis E. *Lonely Hearts of the Cosmos* (London: Macmillan, 1991)

Rappaport, Rhoda. *When Geologists were Historians 1665–1750* (Ithaca, New York & London: Cornell University Press, 1997)

Roger, Jacques. *Buffon: a Life in Natural History*, trans. Sarah Bonnefoi (Ithaca & London: Cornell University Press, 1997)

Rossi, Paolo. *The Dark Abyss of Time* (Chicago: University of Chicago Press, 1984)

Rudwick, Martin J. S. *The Meaning of Fossils* (London: Macdonald, 1972)

Schneer, Cecil J. (ed.). *Toward a History of Geology* (Cambridge, Mass.: MIT Press, 1967)

Ussher, James. *The Whole Works . . . of James Ussher. With a Life of the Author, and an Account of his Writings*, 17 vols, ed. Charles R. Elrington (Dublin 1829–64)

——*The Annals of the World. Deduced from the Origin of Time* . . . (London: E. Tyler, for J. Cook at the sign of the Ship in St Paul's Churchyard, and for G. Bedell, at the Middle-Temple-Gate in Fleet Street, 1658)

Westfall, Richard. *Never at Rest, a Biography of Isaac Newton* (Cambridge: Cambridge University Press, 1980)

Wilson, David. *Rutherford, Simple Genius* (London: Hodder and Stoughton, 1983)

Wilson, Leonard G. *Charles Lyell: the Years to 1841: the Revolution in Geology* (New Haven & London: Yale University Press, 1972)

Acknowledgements

I would like to thank the numerous people who have answered my questions and phone calls over the last two years. In particular the large number of astronomers who so generously gave up their time to talk to me: Allan Sandage, Gustav Tammann, Virginia Trimble, Wendy Freedman and Jeremy Mould for supplying information on the long-running hunt for the Hubble constant; Saul Perlmutter, Gerson Goldhaber, Isobel Hook, Rob Knop and Peter Nugent, of the Supernova Cosmology Project team, and Brian Schmidt of the High-Z team for details of their search for the deceleration of the universe; Neta Bahcall and Erik Leach for briefing me on the latest developments in cosmology; and finally Charles Lineweaver for explaining how he combined other astronomers' results to calculate the age of the universe.

I am indebted to Lorna Patterson for kindly sending me information about the work of her late husband, the Caltech geochemist Clair Patterson. His discovery marked the end of a long quest for the age of the earth. For details of that search I must thank the many historians who have mined these seams before me and brought the choicest nuggets to the surface. In addition I would like to thank the numerous librarians and archivists who have assisted with my research: the helpful staff at the Science Museum Library in London; Christina Mackwell at Lambeth Palace Library; the staff of the British Library; and Adam Perkins at the Darwin Archive at

Cambridge University Library. I very much enjoyed my visit to the Bodleian Library in Oxford where, to gain entrance, I had to promise that I would not 'kindle flame' in the reading room, and where Ussher's eyes, in a seventeenth-century portrait hanging high on the wall, seemed to follow my every move.

Finally I would like to thank Ken and Carole Burton, Tricia Bowker and Raimondo Aiello for supplying translations at short notice; Patrick Walsh, my agent, for his tireless energy and enthusiastic support; Clive Priddle, my editor at Fourth Estate, for his well judged and invaluable editorial notes, and Kate Balmforth and Steve Cox for taking such care with the text. Last, I would like to offer a special thanks to Tim Boon for his discerning comments and helpful suggestions on the typescript.

Picture Credits

Line illustrations on pp. 131, 199, 227, 268 by Miles Smith-Morris. Pages 9, 46, 59, 92, 133, 153 Mary Evans Picture Library; p. 45 Castello del Buonconsiglio, Monumenti e collezioni provinciali, Trento; p. 64 British Library (Burnet, *Sacred Theory of the Earth*); p. 74 National Portrait Gallery (mezzotint by William Humphrey after unknown artist); p. 80 British Library (Scheuchzer, *Homo diluvii testis*); p. 89 Photo RMN – G. Blot/J. Schor (painting by F.-H. Drouais); p. 117 British Library (Borelli, *Historia et Meteorologia Incendii Aetniaei*); p. 119 British Museum (mezzotint by William Ward after a painting by Andrew Geddes); p. 135 National Portrait Gallery (lithograph by C. Hullmandel after George Rowe); p. 157 Cambridge University Library; p. 174 Ann Ronan Picture Library; p. 191 Rutherford Museum, McGill University; p. 202 Harvard College Observatory; p. 205 Henry Huntingdon Library, San Marino, California; p. 209 Royal Astronomical Society (Hale Observatories); p. 219 American Astronomical Society (adapted from the *Astrophysical Journal*); p. 224 Archives Lemaître, Université Catholique de Louvain, Institut d'Astronomie et de Géophysique G. Lemaître, Louvain-la-Neuve; p. 248 California Institute of Technology.

Index

Figures in italics indicate captions.

Autolycus 13–14
autumn equinox 35, 36, 84
Auvergne, France 137–8, 140, 142

Baade, Walter 230–32, 253
Babylon/Babylonians 4, 27, 31
Bacon, Francis 17
Bahcall, Neta 270
Bainbridge, John 27, 28
Banqueting Hall, Whitehall 8, 37
Barwick, Peter 74
Beagle, HMS 146, 147–9, 151
'Beaumont' (an English Jesuit) 19
Becquerel, Henri 177, 178
Bede 14
Bell Telephone Laboratory,
    Holmdel 250
Bentley, Richard 85
Bernard, Nicholas 30
beta particles 179, 180
Bible
  authenticity 21
  the beginning of the universe
    239
  chronology and geological time
    121
  Creation story 5, 7
  discrepancies 41
  genealogy 11–12
  geology and 143, 144
  Guy prints under licence 38–9
  Hebrew 20, 21, 29, 46, 50
  Horner's critique 162
  literal interpretation of 53
  Newton defends its chronology
    83
  Septuagint 20, 29, 47
  two chronologies 52
  Ussher's calculation in 1, 10, 38,
    39, 162, 163, 173
  view of the universe 10
  and voyages of exploration 40
  see also New Testament; Old
    Testament

Big Bang 226, 233, 241–3, 255,
    256, 272–5
  remnants of 251
Big Crunch 256
biology 129, 190, 236
Black, Joseph 122
blue-shift 212, 213
Bolshevik revolution (1917) 239
Boltwood, Bertram Borden
    190–93, 245
Bondi, Hermann 240–41
Bradley, James 200
Brahma 4
British Association 167, 181
British Broadcasting Corporation
    (BBC) 241
British Museum, London 132, 162,
    185
Brongniart, Alexandre 133
Bronstein, Matvei 240
Brown, Harrison 246
Browne, Sir Thomas 29, 53
Bruno, Giordano 54
Brydone, Patrick 115–20, *119*, 140
  Tour 119–21
Buckland, Rev. William 134–5,
    135, 138, 139, 140, 142–5
Buffon, Georges-Louis Leclerc de
    88–114, *89*, 120, 132, 168,
    244
  borrows ideas from de Maillet
    103
  cooling earth experiments
    107–10
  date for the beginning of the
    world 106, 110, 112–13, 114
  director of the *Jardin du Roi* 88,
    93, 104
  and the earth's hot core 106
  forced to retract over *Histoire
    naturelle* 99–100, 111
  and Newton 91, 94, 98, 105–6
  planetary movement 94–5
  science park at Montbard 91

Index

Index

Index

Index

Index

Index

Index